Signals and Communication Technology

Series Editors

Emre Celebi ⓘ, Department of Computer Science, University of Central Arkansas, Conway, AR, USA

Jingdong Chen, Northwestern Polytechnical University, Xi'an, China

E. S. Gopi, Department of Electronics and Communication Engineering, National Institute of Technology, Tiruchirappalli, Tamil Nadu, India

Amy Neustein, Linguistic Technology Systems, Fort Lee, NJ, USA

Antonio Liotta, University of Bolzano, Bolzano, Italy

Mario Di Mauro, University of Salerno, Salerno, Italy

This series is devoted to fundamentals and applications of modern methods of signal processing and cutting-edge communication technologies. The main topics are information and signal theory, acoustical signal processing, image processing and multimedia systems, mobile and wireless communications, and computer and communication networks. Volumes in the series address researchers in academia and industrial R&D departments. The series is application-oriented. The level of presentation of each individual volume, however, depends on the subject and can range from practical to scientific.

Indexing: All books in "Signals and Communication Technology" are indexed by Scopus and zbMATH

For general information about this book series, comments or suggestions, please contact Mary James at mary.james@springer.com or Ramesh Nath Premnath at ramesh.premnath@springer.com.

Sanja Bauk
Editor

Maritime Cybersecurity

Springer

Editor
Sanja Bauk
Estonian Maritime Academy
Tallin University of Technology
Tallinn, Estonia

ISSN 1860-4862   ISSN 1860-4870   (electronic)
Signals and Communication Technology
ISBN 978-3-031-87289-1   ISBN 978-3-031-87290-7   (eBook)
https://doi.org/10.1007/978-3-031-87290-7

This work was supported by Tallinn University.

© The Editor(s) (if applicable) and The Author(s), under exclusive license to Springer Nature Switzerland AG 2025. This book is an open access publication.

Open Access This book is licensed under the terms of the Creative Commons Attribution-NonCommercial-NoDerivatives 4.0 International License (http://creativecommons.org/licenses/by-nc-nd/4.0/), which permits any noncommercial use, sharing, distribution and reproduction in any medium or format, as long as you give appropriate credit to the original author(s) and the source, provide a link to the Creative Commons license and indicate if you modified the licensed material. You do not have permission under this license to share adapted material derived from this book or parts of it.

The images or other third party material in this book are included in the book's Creative Commons license, unless indicated otherwise in a credit line to the material. If material is not included in the book's Creative Commons license and your intended use is not permitted by statutory regulation or exceeds the permitted use, you will need to obtain permission directly from the copyright holder.

This work is subject to copyright. All commercial rights are reserved by the author(s), whether the whole or part of the material is concerned, specifically the rights of translation, reprinting, reuse of illustrations, recitation, broadcasting, reproduction on microfilms or in any other physical way, and transmission or information storage and retrieval, electronic adaptation, computer software, or by similar or dissimilar methodology now known or hereafter developed. Regarding these commercial rights a non-exclusive license has been granted to the publisher.

The use of general descriptive names, registered names, trademarks, service marks, etc. in this publication does not imply, even in the absence of a specific statement, that such names are exempt from the relevant protective laws and regulations and therefore free for general use.

The publisher, the authors and the editors are safe to assume that the advice and information in this book are believed to be true and accurate at the date of publication. Neither the publisher nor the authors or the editors give a warranty, expressed or implied, with respect to the material contained herein or for any errors or omissions that may have been made. The publisher remains neutral with regard to jurisdictional claims in published maps and institutional affiliations.

This Springer imprint is published by the registered company Springer Nature Switzerland AG
The registered company address is: Gewerbestrasse 11, 6330 Cham, Switzerland

If disposing of this product, please recycle the paper.

Preface

The sea is a vital ecosystem for life on our planet. It produces up to half of the world's oxygen, absorbs carbon dioxide from the atmosphere, provides nearly three billion people with essential proteins, and controls the planet's temperature. It also provides a variety of other resources used by humans. Maritime transport is the safest and most effective way to move goods and raw materials around the world; therefore, it facilitates around 80% of global trade. However, significant pressures from climate change, ocean acidification, rising sea levels, fluctuating fish stocks, natural and man-made disasters, and other factors threaten the sustainability of marine habitats and coastal communities. Furthermore, the maritime industry and businesses face the challenge of their relative lack of digitalization compared to other economic sectors. The aviation industry, for instance, is much further ahead in terms of digitalization. Namely, there are millions of registered drones, but very few autonomous research, passenger, cargo, and military vessels.

What could the reasons for this be? Maritime is older than aviation and more conservative. The regulations in maritime are not unique, and even where they are in place, they are not always adhered to. While almost all countries have signed the Safety of Life at Sea (SOLAS) Convention, there are still many non-SOLAS vessels, such as fishing boats, pleasure yachts, and small cargo vessels. Additionally, there is intense competition in the shipping sector, with some participants reluctant to publicly disclose information essential to their successful performance. Digitalization also brings with it the risk of cyber-attacks. This could explain, to a certain extent, the slow pace of digitization in shipping and port logistics on a global scale.

However, the acceleration of digitization in the maritime sector began with the commercialization of the Internet. In the early 1990s, ECDIS (Electronic Chart Display and Information System) appeared on ships as part of the IBS (Integrated Bridge System). As navigation becomes more digital, so too does the ship's propulsion system, which is increasingly being driven and controlled digitally. The navigation bridge and the ship's mechanical complex are becoming more and more like computer centers for collecting, processing, storing, and displaying information to ensure safe, effective, and efficient navigation. In a relatively short time, a diversification has emerged between IT (Information Technology) and OT

(Operational Technology), but we are now moving toward their convergence. These processes are monitored by land-based VTS (Vessel Traffic Service) centers, as well as by advanced satellite CNS (Communication, Navigation, and Surveillance) systems.

In some ports, administrative tasks and operational processes are being digitized as physical and computer systems increasingly converge. Major ports today use EDI (Electronic Data Interchange), PCS (Port Community System), SW (Single Window), and ELM (Electronic Logistics Marketplace). The Internet of Everything (IoE) emerged in the early 2010s, further connecting the virtual and real worlds, while enabling micro- and macro-control and management simultaneously. Programs and data are being transferred from data centers on ships, in ports, and at traffic control centers to the Cloud. Work is underway to develop cost-effective underwater data collection and processing centers. Furthermore, the volume of data collected in real time is growing exponentially, giving rise to a new field of study known as Big Data. Big Data provides a more detailed and comprehensive insight into real-world processes, including those in the maritime. This new discipline will breathe new life into digital twins of ships and ports as well. This has been followed by the development of automated expert systems based on deep and federated machine learning and artificial intelligence algorithms.

The maritime digitization corps also includes surface and underwater vessels with varying degrees of autonomy. These vessels can be remotely controlled, and some can operate completely autonomously for a period. Most of those vessels cooperate with air assets, such as drones and low-orbit pseudo-satellites.

Obviously, digitalization in shipping and port management is advancing in several areas, with an increasing emphasis placed on ensuring the safe operation of complex systems. As digitalization increases, so does the risk of safety and security issues, including cyber-attacks. Consequently, efforts are being made to detect and prevent such risks. The rapid pace of digitalization must be considered in the context of technical, legal, ethical, and societal needs and preferences. This collection of studies aims to raise readers' awareness of the potential risks associated with cyber-attack vectors in the maritime sector, while also increasing their cyber-awareness and knowledge in technical and human domains. Therefore, I would like to express my gratitude to the fellow contributors for their hard work and dedication to this relatively new applied science discipline, which will become one of the top priorities in the coming decades.

Tallinn, Estonia Sanja Bauk
2024

Acknowledgments

This research was funded by the EU Horizon 2020 project 952360-MariCybERA.
I appreciate the co-authors' contributions to this publication.

Contents

The Future of Cybersecurity at Sea: Human vs Human or AI vs AI? 1
Anatoli Alop

Addressing Maritime Workforce Cybersecurity Skills Development 15
Júlia Grosschmid

Rethinking Seafarer Training for the Digital Age 29
Dan Heering

Ports of Tallinn and Koper Comparative Analysis Including
Cybersecurity 55
Sanja Bauk, Bojan Beskovnik, and Seçil Gülmez

Simulating Cyber-Attacks on the Unmanned Sea-Surface Vessel's
Rudder Controller 83
Igor Astrov and Sanja Bauk

A Scope Review of Secure Broadcasting Protocols for the
Automatic Identification System 103
Leonidas Tsiopoulos and Risto Vaarandi

Using Incremental Inductive Logic Programming for Learning
Spoofing Attacks on Maritime Automatic Identification System Data 123
Aboubaker Seddiq Benterki, Gabor Visky, Jüri Vain,
and Leonidas Tsiopoulos

Technical Considerations for Open-Source Intrusion Detection
System Integration in Marine Vehicles 143
Gabor Visky, Dariana Khisteva, and Olaf Maennel

Enhancing Cybersecurity in Marine Vessels: Integrating
Artificial Neural Networks with Inertial Navigation Systems for
Resilience Against GPS Cyber-Attacks 161
Yiğit Gülmez

A Comprehensive Review of Social Engineering on Maritime Cybersecurity 179
Veera Senthil Kumar Ganesan and Mihir Chandra

Toward Secure Marine Navigation: A Deep Learning Framework for Radar Network Attack Detection 195
Md. Alamgir Hossain, Md. Delwar Hossain, Latifur Khan, Hideya Ochiai, Md. Saiful Islam, and Youki Kadobayashi

Improving Security and Privacy with Raspberry Pi Devices 217
Radoje Džankić, Zvonko Bulatović, and Sanja Bauk

Cybersecurity and Commercial Shipping: Is There a Need for Unification? 235
Ann Fenech and Sebastien Lootgieter

Index 247

The Future of Cybersecurity at Sea: Human vs Human or AI vs AI?

Anatoli Alop

1 Introduction

All of us are witnesses to the fact that in recent decades, new information technologies have begun to penetrate our lives more and more rapidly and, in essence, forcefully determine further developments in all areas. Artificial intelligence (AI) is arguably the fastest growing technology in the world today [1]. The market size of the artificial intelligence industry is growing exponentially every year, and forecasts show exponential growth of the market in the future as well (Fig. 1).

Maritime is no exception. However, following the developments and trends, one may get the feeling that somehow overly optimistic expectations and sometimes even euphoria is taking place regarding digitalization. More and more people are starting to say here and there that soon artificial intelligence (AI) will take over everything and immediately all the worries of both humanity and individuals will be solved, including in maritime— the level of maritime safety will immediately jump to an unprecedented level, because such a source of making stupid mistakes as human being will be removed from seafaring and the decision-making process; maritime transport becomes nature's best friend, because a smart artificial intelligence can invent such ships, fuel types, navigation methods, supply chains and everything else and put them to work, so that the negative impact on the marine environment starts to decrease and reaches zero; digitalization will be every shipowner's dream come true, as those who adopts AI will be rewarded with significant economic gains compared to those who don't, etc.

Of course, these expectations are partially justified, and the widespread application of AI in maritime industry will certainly bring many positive developments.

A. Alop (✉)
Estonian Maritime Academy, Tallinn University of Technology, Tallinn, Estonia
e-mail: anatoli.alop@taltech.ee

© The Author(s) 2025
S. Bauk (ed.), *Maritime Cybersecurity*, Signals and Communication Technology,
https://doi.org/10.1007/978-3-031-87290-7_1

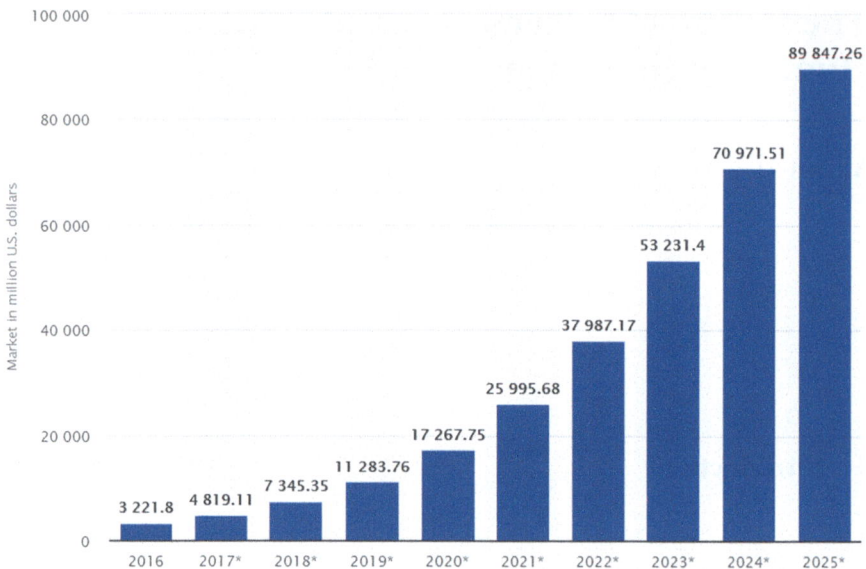

Fig. 1 The growth of the AI industry market size (2016–2025) in US million dollars. (Adopted from [1])

At the same time, one must not lose one's common sense and critical mind when analyzing and predicting future trends. The digitalization of all areas of human activity is characterized by several accompanying problems, which are waiting to be solved effectively in one way or another; one of them is the issue of ensuring cybersecurity. This problem is of course not only relevant for maritime, but now for all areas of our life, because there is no longer any area that is far from digitalization. Actually, we can't say that the problem of cybersecurity at sea is somehow very specific compared to all other areas of digitalization of human activity. If to look for something important specifically for maritime, the negative effects related to the realization of this problem may have sometimes more severe consequences at sea than anywhere else. Therefore, although this chapter tries to look at the problem in the context of maritime, above all shipping cybersecurity, the author's reflections and the conclusions drawn from them are unintentionally transferring to cybersecurity more broadly. Namely, the main question that is being tried to be answered is whether and if, how can the increasingly widespread use of AI affect cybersecurity and what can be its further developments in this field? Is there any hope that these effects will be positive only? Or does it not depend on the rate and nature of the implementation of AI on ships, and the fight against cyber piracy remains the same activity with difficult to predict results as it unfortunately appears to be now? There is hardly anyone anywhere who would answer these questions in the affirmative or in the negative with great certainty and conviction. The author of this chapter also does not pretend to this but tries to offer his thoughts on this topic to the reader in the more or less readable form.

2 Problem Statement

Today, cybersecurity has undoubtedly become one of the most important and challenging problems in any kind of human activity, including maritime. Cybercrime, like the adoption of AI, is also on the rise; only reported cybercrime losses have grown strongly over the past decade, with exponential growth clearly visible in 2020–2022 (Fig. 2).

As has often happened in the history of mankind, every initially positive and forward-looking development is accompanied by negative effects and consequences caused by it or directed toward it. Some of them are caused by natural forces, laws of physics, and other objective factors, but a large part of them are still human-made ones. Their reason is mostly people's not the most pleasant natural ore acquired during life, especially during childhood, personal characteristics such as greed, fear, pettiness, hatred, inferiority complexes, etc. There is always a certain, sometimes even not very small number of people who try to use the achievements of techniques and technologies to realize their personal, often illegal and even criminal intentions—to get rich at the expense of other people, in one way or another to gain power over other people and groups of people, etc. Considering the extent of the global adoption of the Internet and digital technologies in recent decades, it is not surprising that the number and severity of cybercrimes have reached unprecedented levels in recent years and their growth trend continues. The author leaves aside aspects related to real wars and geopolitical instability and deals with what happens in "world of peace." The expression "world of peace" has been put in quotation marks, because it is difficult to call the struggle taking place in the field of digital

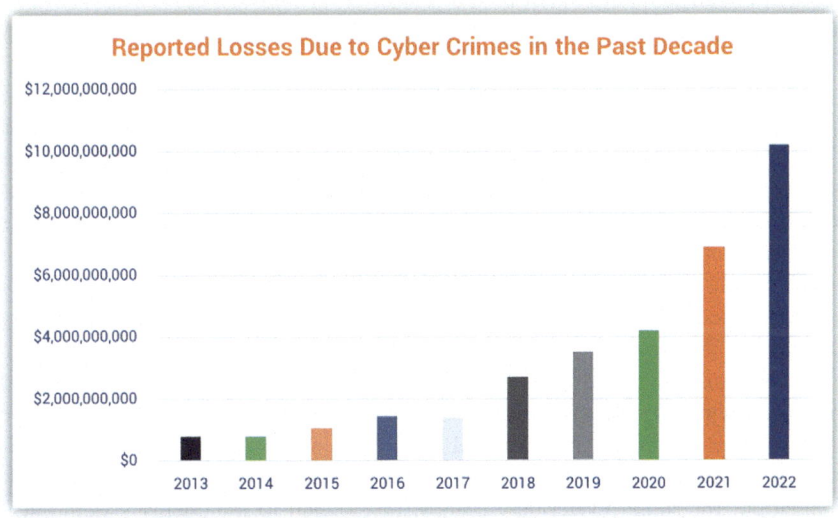

Fig. 2 Growth of cybercrimes in the period 2013–2022. (Adopted from [2])

technologies something calm and positive, on the contrary, the words "cyber war" perhaps best characterizes what is happening.

In this war, there is no front line, rather it runs everywhere and nowhere, there is not clearly defined and unambiguously identified enemy, it is not governed by internationally recognized conventions of warfare, and as sad as it sounds, in this war it is not and is difficult to expect to be one and indisputable the winner. The battles of this war take place all the time, now it can be said that every day and all over the world, the opponents in this war are inventing and deploying more and more new "weapons systems," be they defensive or offensive systems, that is, there is a genuine "arms race," and accordingly sometimes one side wins battle, sometimes the other. It may be said that it is a war of attrition, to which there is no end in sight, whatever it may be, and in which it is difficult to hope that there will ever come a moment when one side has won indisputably and definitively. Until recent years, it has been a war with people fighting on both sides, let say the "good" and the "bad" guys. They both sides have recently started to increasingly use smart computer programs as assistants, eventually what is called artificial intelligence (AI), but so far, they have mainly played the role of command executors and technical aids, at least people hoped so. Recently, however, the recognition of AI as an intellect, i.e., attributing to it some, if not many, characteristics and abilities that are only characteristic of humans until now is taking place more and more. This is inevitably accompanied by heightened expectations to the point that AI will soon start to live (or to some extent already living?) a completely independent life, make completely independent decisions, and in general successfully take over the mission of advance the economy and even all the humankind prosperity. Some praise AI to the skies and are sure that soon it will take over everything and eternal prosperity and happiness for all will come, some see these perspectives in much darker shades and are pessimistic about an AI-driven future. In the author's opinion, both tend a little to extremes, the reason being once again the emotions and beliefs that mainly govern people's attitudes and imaginations.

In the introduction, it was mentioned that in recent years, the use of AI in the commission of cybercrimes has been growing at a tremendous speed. Attackers use AI to create increasingly more perfect, complex, and effective cyber threats and use them to achieve their goals. Even the best and highly qualified security teams in companies and institutions are no longer able to successfully resist such attacks using traditional security measures and defense mechanisms, so it is self-evident to use AI on the defense side as well. Undoubtedly, a security system supported or even controlled by AI can perform its tasks significantly more effectively, including the fact that AI is able to act proactively in many cases, rather than reacting to threats only after they are discovered, which is common for systems managed and controlled by people. But the question is becoming more and more relevant: can we still completely and fully trust artificial intelligence with our digital security?

Now, there is no yet serious talk about the possibility that the AI itself can become "malicious" for some mysterious reasons, i.e., switch to the enemy's side, but it is also possible to hack into the AI itself and turn it against the user, producing incorrect or dangerous instructions, whereas another AI system can of course be used for this purpose. Creators and developers of security systems based on AI naturally recognize the existence of such threats and offer instructions such as "use AI as a helper, not a replacement for human knowledge," "keep sensitive details out of AI systems," or even "implement systems for human oversight and intervention [3]," but this means a significant reduction in the effectiveness of the security system, because every system is only as weak as its weakest link. By involving a person in the system as a watcher and controller, the effectiveness of the system inevitably comes down to the limited capabilities and possibilities of the person. Another option is, of course, that to eliminate humans from the process, the AI-controlled security system itself is secured by another AI-based security system, but then according to the idea, the latter would also need a duplicative security system. In this way, absurdity can be reached rather quickly, and such duplication and cross-security do not seem to be very attractive and economically feasible.

So, the problem is that a human is less and less able to adequately control the AI's activity, especially where one AI attacks and another AI defends—the interaction (otherwise, the fight) between them takes place so quickly and using such large amounts of data and algorithms that even the identification of need for human intervention and even more so the timely implementation of the necessary measures is already now clearly beyond human power, and it is an ever-increasing trend. It cannot be said that the developers and users of AI as a cybersecurity system do not understand this. M. Oxenwald [3] says: "Evaluate the AI's development process and underlying algorithms for potential biases or ethical concerns." This can even be accepted in theory, but how to bring biases or ethical issues into play in the context of AI "behavior" and developments and whether this is possible at all remains a question for the author of this chapter, to which finding an adequate answer may be difficult if not impossible, at least now. In the next section, we will try to see if and how it would be possible to make an AI "think" and decide like a human, i.e., what are the human qualities that allow a person to distinguish between bad and good and act accordingly, and whether there is hope that AI will one day be able to do this as well.

3 Human and AI in Interaction as Cybersecurity Subjects

As explained in the previous section, the battle in the field of cybersecurity is moving more and more on the AIs' battlefield. If until recently human has been more in the role of the creator, commander, and controller of AI or at least of computer algorithms, then soon the center of gravity must inevitably shift to partnership relations between human and AI. A real and functioning partnership, however, requires trust between partners, considering trust in a human context. Currently, we use the term "reliable" also for machines, devices, programs, forecasts, etc., but by this we mean "working condition," "reliability," "provability" rather than trust in the literal sense of the word. However, trust between people means that we believe that our partner, friend, or whoever has the same beliefs and values as us and will not let us down in any case, without us constantly having to input or remind them of anything. In fact, humans should develop a similar relationship with AI, where we must be convinced that even in difficult and/or unforeseen or critical situations, AI will be able find its own solutions that may not have been pre-programmed by us and, more importantly, that these solutions are correct, perhaps suitable for us and are the best in the given situation. It should be mentioned here that when saying "we," the author means the persons on the side of goodness. As we see, such human concepts as "goodness" and "badness," "good guys," and "bad guys" have come into play. Next, we will focus, among other things, on defining these terms, including in the context of the inclusion of artificial intelligence.

3.1 The Psychology of Decision-Making

When we talk about the fight against cybercrime, we need to talk about the decision-making process, because it is critical to make the right decisions quickly and then implement the most effective measures just as quickly when countering cyber-attacks. I emphasize that the decision must first be *right*, because the high speed of making a wrong decision has a negative rather than a positive effect. In any case, maybe not so quickly made, but the only right decision is better than a wrong decision made quickly and then implemented just as quickly, the latter will most likely only lead to negative consequences.

Let's look at what factors influence people's behavior and ultimately largely determine what decisions people make. According to Rassmussen [4], human behavior and thus decision-making are formed on three levels: knowledge-based, rule-based, and skills-based. To behave correctly, that is to make the right decisions, preparatory work is the most important at all these levels. Let's note that we are now talking about the right and wrong decisions made by the "good" ones, that is, we are not talking about malice or crime-caused decisions. In this variant, where it is necessary to choose the right behavior, including the right means and methods of countering a cyber-attack, the advantage of AI over humans is indisputable. According to such characteristics as the speed of gathering, analyzing, and acting on information, as well as the ability to learn more, AI has a mountain height advantage over human. Moreover, AI never gets tired, its attention is always in focus, and it is never disturbed by any moods or negative emotions. So, in standard situations, i.e., those that are foreseen by AI algorithm or its algorithm allows to learn more all the time, as well as, for example, possible cyber-attacks have been played previously out on simulators, the AI is up to its tasks. It does not have to decide whether its action is "good" or "bad," it is always on the side of good for it a priori.

However, everything is not so clear-cut when the situation is uncertain, and right choice must be made to make decisions. In this case, the process usually includes several stages: identification of the decision, gathering the necessary information, evaluating the options, making a choice, and implementing the decision [5]. The first stage involves realizing that a decision needs to be made; it is based on the identification of a process or situation that requires a decision, such as a cyber-attack, for instance. The second step, gathering information, is essential for making informed choices; sources of information can be, in addition to what was learned before, personal experiences, advice from others, or data obtained from the Internet or even from various media channels. This applies to humans, but AI also gathers information in a similar way, depending on what sources it has access to. The third stage, or evaluating the options, requires an assessment of the possible outcomes and consequences associated with each alternative; at this stage, psychological factors come into play that influence people to give pros or cons to different options. These factors are emotions, past experiences, and cognitive biases that make the evaluation of options subjective. If in the case of AI, we can talk about past experiences mainly with the algorithm in mind, then AI doesn't have emotions and biases, at least using the human dimension; in fact, this cannot necessarily be considered as shortcoming of AI in standard or protocol-prescribed situations, where emotions and biases would rather create confusion in the AI acting. The fourth stage, or making a choice, means choosing one option from among several alternatives; this stage is also affected by various psychological factors for people, such as risk tolerance, emotional state, and personal values, these factors probably do not affect the choice of AI as well.

3.2 "Good" People and "Bad" People

Everything would be fine if all people behaved according to laws, generally accepted rules and moral and ethical principles. In that case, there probably wouldn't be those trying to do something against the laws and rules, but then there wouldn't be a reason to write this chapter either. Unfortunately, this is not the case, and as written above, the "cyber war" is going on and is merciless and harsh; in this war, AI will play an increasingly decisive role on both sides, until eventually it will take over all control and decision-making. As it was also said, human intervention in it and achieving even some kind of control over the processes have a big question mark and will most likely prove to be impossible. The related question is—is AI capable of making difficult decisions on its own, including in situations where there are several choices and where, figuratively speaking, it is necessary to be able to distinguish between good and bad?

Let's first look and try to understand why some people are good and some are bad? Rather, what are the factors and circumstances that keep most people from killing or stealing; the most people are law-abiding citizens and loving and loved family members. Perhaps the next question is, where do people get their moral and ethical principles, beliefs, correct convictions, and values? The answer is obvious—they get them over from their parents and other family members, teachers at school, good books, and movies, but nowadays sometimes also from the Internet. It is customary to call all this with two words "children's room." But then the question arises—why aren't all the people like that? All children have more or less access to all these sources of information, so why do some go down the wrong path? Because just finding and collecting of information are not enough, the most important thing is to give the right assessments and make the right choices on base of them. Loving parents, kindergarten and schoolteachers, good friends, and others help to make the right choices. But there are other influences, for example, movie heroes, models, bands, and other youth idols, whose effects are not always positive; the internet isn't always the best indicator of the right path either, don't speak about social media. Not coincidentally, a law restricting the use of social media by young people is currently to process in Australia [6]. So not for everyone becoming an adult might go very well. But in conclusion, it can still be said that the right "children's room," that is, upbringing, plays a decisive role in what a person becomes. If the upbringing has gone wrong, any additional education will not help.

Why is the existence of a well-established system of values, beliefs, ethical and moral principles of decisive importance for each person, but also for society in general and even for all of humanity? Because without them, it is not possible to make the right decisions in situations where there are no ready-made or tried-and-tested solutions, or it can also be said even about people that these situations are not pre-programmed. In addition to the right upbringing, a normal person has a whole series of purely human qualities that do not allow him to deviate from the right path, even if it may be difficult or unpleasant on a personal level. These are self-esteem, a sense of shame, the need for recognition by society and closed people, empathy, the desire not to sadden or disappoint people whose opinion is very important, etc. These qualities are like a compass that helps you always look and move in the right direction. For those for whom this compass does not work well, society has devised indirect or direct instruments of coercion—religions, laws and other legislation, the non-compliance of which is most likely to be followed by punishment, and the like.

3.3 "Good" AI and "Bad" AI?

The reader may ask: what do these reflections on human psychology have to do with the topic of this chapter, which is AI and cybersecurity? The point is that the time has come when we need to find those instruments and solutions that allow us to hand over control and decision-making from humans to computers, or rather to AI, including in the most difficult situations. In fact, AI is going to take over everything very quickly in one way or another, whether we like it or not, the question is whether we watch it happen without fully understanding or being able to intervene, or whether we try to create trustworthy AI systems based on the right values and ethical and moral principles. However, this would mean that AI would have to take over more and more of the characteristics that so far only attributed to humans. AI, of course, doesn't have to do this formally, by collecting relevant information from the Internet and imitating human feelings, but in a real and meaningful way, to understand deeply what is good and what is bad, what is right and what is wrong, why people behave this way or that way. How and whether it is even possible to achieve this is a matter of questions, it seems that at least now it is "mission impossible."

Returning to the role of upbringing in the development of human personality, we ask whether it is possible to apply "upbringing methods" to AI? Not really, it seems, even the question itself seems silly. The process of developing a human personality, including raising a child, takes decades, during which a person develops the right values, beliefs, and principles, largely thanks to the example of other people, first parents, but also to other influencing factors. After all, the child is often simply told that "no one does that," "this is not good manner," "you will embarrass yourself," etc., after many repetitions, and seeing that others do not do it, the child also develops the right attitude and understanding. Can we imagine AI being taught that one or another action is undignified or unethical? In principle, we can regarding some specific things, but the need is rather that AI, like a human, must be able to generalize and draw correct conclusions even in unknown situations or activities. Or is it possible for an AI to take an example from someone? Theoretically, again, it is, but what role models can AI find on the internet? Most of them will certainly not be the most positive. So, a "children's room" for AI is currently more of a science fiction topic than something realistically possible.

We can imagine AI as a child a few months old, who does not know yet what is good and what is bad, and who is essentially not held back by anything; the development of the best human qualities would take years and years, and that only in the right "children's room." However, unlike a human child, AI in "baby shoes" has enormous power and abilities with which it can accomplish great deeds, including, unfortunately, crushing and demolition, because it lacks moral and ethical brakes. We can only hope that AI will also be able to develop these qualities, or rather, it is inevitable if we finally want to see an AI that only makes the right decisions. However, time is critical, because unlike ethics and morality, the technical capabilities of AI are growing at a dangerously high rate, especially since some AI "parents," i.e., creators and users of programs with criminal tendencies, are happy to exploit AI capabilities for very bad purposes.

3.4 Regulations for AI?

Since the idea of AI adopting ethical and moral principles and forming values does not seem to be very realistic, we turn to the possibility of regulating AI's activities by establishing certain regulations. It should be mentioned here that we are not talking about written laws or other conventional legislation made mandatory for compliance at the level of countries and the international community. The reason is simple, such legal regulations would be intended for people, but there is nothing new in this, all countries have many laws that already regulate people's activities and behavior. If everyone and everywhere fulfilled them, then there would be no one who uses AI to organize illegal activities, including cyber-attacks. We are talking about such rules that could be adopted by AI itself, that is, included in one way or another in AI algorithms and thus prevent the illegal activities of AI itself or its users. These would be certain imperatives that no AI could or should violate, i.e., AI would always refuse in one way or another from contributing to illegal, unethical, etc. activities, the more so directly committing them. This means that they should be the most important part of the core AI algorithm, which determines the operation of the entire algorithm. Again, the same question, is such an option even possible? Unfortunately, it seems that not much either. This would require complete and total control over the creation and use of AI programs on a global scale. If you even imagine that some mystical "world government" could handle this, the probability of such a government emerging is zero. Besides, it's too late, this genie is out of the bottle, and no one is going to be able to put it back in.

The first attempt to establish some rules of behavior for robots, or actually for AI, was made by the American science fiction writer Isaac Asimov in his short story "Runaround" in 1942, which he named the "three laws of robotics" [7, 8]. With them, Asimov tried to create an ethical system for humans and robots. The laws are as follows: "(1) a robot may not injure a human being or, through inaction, allow a human being to come to harm; (2) a robot must obey the orders given it by human beings except where such orders would conflict with the First Law; (3) a robot must protect its own existence as long as such protection does not conflict with the First or Second Law." Later, Asimov added another rule, known as the fourth or zeroth law, which takes precedence over the other three. It sounded like this "a robot may not harm humanity, or, by inaction, allow humanity to come to harm." It is clear that in today's world, which is essentially already full of robots, there is not the slightest possibility that robots could even fulfill any such rules anywhere, firstly because most robots (e.g., several kinds of drones, but not only) are developed and used by military structures; their nature and use are in direct contradiction to Asimov's laws. In fact, these laws cannot be used as a basis for any regulation today, we bring them here as an example of the first attempt to somehow reconcile ethical principles with the world of robots or AI. Criticisms of these laws can also be read in source [9].

Apart from the mentioned reasons, which do not allow the application of Asimov's laws today, there would be great difficulties in their effective use even in modern civilian life, considering the problem of cybersecurity. The "human" mentioned in the first three laws is an abstract concept, i.e., all living beings with the physical characteristics of humans would fall under it, regardless of their qualities, i.e., basic values, ethical and moral principles, intentions, etc., that is, the rules do not distinguish between "good" and "bad," rather they assume by default, that all people are good. We know that this is not the case, and one would have to be very naive to hope that it ever will be. Even more general and vague is the subject of the fourth or zeroth law, which is humanity as a whole; humanity cannot be viewed in this context as a whole, it consists of individual human beings, who again can be both good and bad, danger for other people or not and so on.

Unresolvable dilemmas are easy to arise. Let's imagine a situation where, for example, a cybercriminal deliberately uses the AI programmed by him to carry out a cyber-attack on an autonomous tanker, as a result of which this tanker collides with a manned ship; therefore, some people are killed, some are injured, in addition, a large marine pollution occurs, causing harm to many people and to humanity as a whole. By carrying out a criminal order, the AI violates the first, second, and zeroth laws. The dilemma is to refuse to fulfill it and to pass on the necessary information to the authorities, which will allow them to arrest the criminal. Let's say this happens in a country where such a crime is punishable by death, in this way the AI is again breaking the first law. It seems that getting these laws to work effectively would be an unrealistic task.

We can imagine a situation at sea that is not directly related to cybercrime, but still requires difficult choices to be made. We are talking about an AI-controlled autonomous ship again. On a waterway with heavy traffic, a situation may arise where a ship, complying with the COLREGs rules, puts itself and other ships at greater risk; this happens, for example, when a ship, which according to the COLREGs rules is obliged to give way, does not do so and there is a real risk of collision between ships. In this case, COLREGs allows and, from a certain moment, demands violation of the rules, i.e., giving way to a ship that should have given way. However, it may happen that a collision can only be avoided by steering the ship aground (for this, of course, the seabed in the area must allow it to be done without major breakages and damages). But already here the problem of a difficult choice can arise. Presumably, the AI is programmed to protect the interests of the ship owner and the owners of the goods on the ship with all its actions and decisions. Directing the ship aground definitely means acting against their interests, because additional resources of money and time are needed to get the ship off the ground, for possible repairs, and to eliminate possible pollution. The AI can analyze all possible options at its huge speed and find that continuing to drive on the collision course and driving directly into another ship will cause less damage to the owner of the ship it controls. What decision will the AI make?

Situation may be even more complicated. For example, it is an autonomous tanker controlled by AI, while the seabed at the scene is not flat and soft, there may be stones and unevenness, and running aground may lead to extensive marine pollution. But the damage to the ship itself in this case would be less than in the case of continuing the course and colliding. Which turns out to be decisive, the interests of the ship owner or the interests of society in a broader sense? There are no easy answers to these questions. What should the captain of a manned ship be based on when deciding in such a situation? Certainly, on his professional knowledge, experience, good maritime practice, but also largely on what we called above the "compass of ethical and moral principles". Unwittingly, we are back to the fact that AI cannot be allowed to make independent decisions based only on economic expediency, "soft" values and moral and ethical principles must be important influencing factors for AI's decision-making process.

4 Conclusion

Was the question posed in the title answered in this chapter? It must be admitted that not a final and unequivocal answer was found. So to say, cybersecurity formula currently contains too many unknowns for one most effective solution to be found, rather different variants are introduced in different situations and by different players may be used. Which of them will eventually become dominant or prove to be the most effective, it is hard to say now. It is very likely that the situation will continue in the coming years, where there are real people are behind both cyber-attacks and cyber defense, but at some point, they may start to discover that they no longer understand what is happening very well, even more so that they are no longer able to intervene and change something. Depending on how well humans can exploit and/or develop (or let them develop) AI capabilities to achieve their goals, the aforementioned "arms race" will continue in one way or another. The danger that the operation of AI on both sides will become more and more complex and self-developing may lead to the situation that soon no human will be able to keep up with it and human intervention in the processes will become increasingly problematic. In the author's view, one possible (if not the only) way out of such a situation could be the forced introduction of a universal "moral compass" into the programming process of all AI algorithms, although it is far from clear how to develop and implement such programs, even if it is possible at all. Its success or failure will likely determine whether we see more and more powerful and seemingly well-protected systems, or even global networks, collapse in the not-too-distant future, or will humanity manage to "get over the knife edge" this time too.

Acknowledgments Research for this publication was funded by the EU Horizon2020 project 952360-MariCybERA.

References

1. A. Oppermann, *Artificial Intelligence Market Size. DataSeries* (Medium, 2022), https://medium.com/dataseries/artificial-intelligence-market-size-a99e194c184a. Accessed 18 Sept 2024
2. C. Crane, A look at 30 key cyber crime statistics [2023 data update], Hashed Out by the SSL Store T.M (2023), https://www.thesslstore.com/blog/cyber-crime-statistics/. Accessed 18 Sept 2024
3. M. Oxenwaldt, *AI Fighting AI: The Future of Cybersecurity—Are You Ready?* (Columbus, 2024), https://www.columbusglobal.com/en/blog/ai-fighting-ai-the-future-of-cybersecurity. Accessed 18 Sept 2024
4. J. Rasmussen, Human errors: taxonomy for describing human malfunction in industrial installations. J. Occup. Accid. **4**, 311–333 (1982)
5. K. Cherry, *The Psychology of Decision-Making Strategies* (Verywellmind, 2022), https://www.verywellmind.com/decision-making-strategies-2795483. Accessed 18 Sept 2024
6. K. Byron, J. Renju, *Australia Plans Social Media Minimum Age Limit, Angering Youth Digital Advocates* (Reuters, 2024), https://www.reuters.com/technology/australia-plans-social-media-ban-children-2024-09-09/. Accessed 18 Sept 2024

7. S.L. Anderson, *Asimov's Three Laws of Robotics and Machine Metaethics* (University of Connecticut, Dept. Of Philosophy, Stamford, CT), https://cdn.aaai.org/Symposia/Fall/2005/FS-43405-06/FS05-06-002.pdf. Accessed 18 Sept 2024
8. A.-S. Brauner, Runaround by Isaac Asimov and the significance of the Three Laws of Robotics in today's world, Seminararbeit, 2015, https://doi.org/10.13140/RG.2.2.16275.14884
9. P.W. Singer, *Isaac Asimov's Laws of Robotics Are Wrong* (Brookings, 2024), https://www.brookings.edu/articles/isaac-asimovs-laws-of-robotics-are-wrong/. Accessed 18 Sept 2024

Open Access This chapter is licensed under the terms of the Creative Commons Attribution-NonCommercial-NoDerivatives 4.0 International License (http://creativecommons.org/licenses/by-nc-nd/4.0/), which permits any noncommercial use, sharing, distribution and reproduction in any medium or format, as long as you give appropriate credit to the original author(s) and the source, provide a link to the Creative Commons license and indicate if you modified the licensed material. You do not have permission under this license to share adapted material derived from this chapter or parts of it.

The images or other third party material in this chapter are included in the chapter's Creative Commons license, unless indicated otherwise in a credit line to the material. If material is not included in the chapter's Creative Commons license and your intended use is not permitted by statutory regulation or exceeds the permitted use, you will need to obtain permission directly from the copyright holder.

Addressing Maritime Workforce Cybersecurity Skills Development

Júlia Grosschmid

1 Introduction

The maritime industry is becoming increasingly digitalised and ever more reliant on digital technologies, making cybersecurity a critical concern. The workforce's role in maintaining cybersecurity has never been more in the spotlight. The safety and security of maritime operations are directly tied to the competence, knowledge, and skills of those working both at sea and on shore, for which seafarers are educated all over the world. From seafarers navigating vessels to shoreside personnel managing complex logistical networks, today every member of the maritime workforce plays a vital role in safeguarding against cyber threats as well. With the rise in digital technologies, the maritime sector is facing an expanding number of cyber risks, where the recent years have seen a marked increase in cybersecurity incidents, exposing vulnerabilities within the whole sector such as vessels, ports, and entire organisations fall victim. The challenges are rooted in the growing complexity of interconnected systems, the slow adoption of new processes, and the potentially shrinking workforce, making cybersecurity skills development an urgent priority.

As the cyber threat landscape continually evolves, so tries the maritime sector's resilience to adjust, yet new threat vectors and opportunities for development will always arise. Some of the new emerging threats in this evolving landscape are the rise of unapproved IoT devices, unsupervised third-party software, and shadow AI—where AI tools and services are utilised without the oversight of essential corporate teams. This phenomenon introduces unpredictable risks that

---

J. Grosschmid (✉)
Tallinn University of Technology, Tallinn, Estonia
e-mail: julia.grosschmid@taltech.ee

can undermine cybersecurity (and safety) efforts, particularly if the workforce is undertrained or unaware of the implications.

As digitalisation continues to reshape the maritime industry ensuring that the workforce is equipped with the necessary cybersecurity awareness and skills are essential for maintaining its resilience and safeguarding the future of maritime operations. The current era can be considered as a transitional period where the active workforce is composed of digital natives—those who effortlessly navigate digital tools and online spaces creatively—and those with a more traditional mindset, relying on acquired knowledge, experience, and verbal communication. Between these extremes lie various degrees of adaptation and traditional and IT proficiency. Intergenerational relationships have evolved, making it essential to reassess how skills development is approached. Digitalisation and interconnectivity brought ease and simplicity; however, it is important not only to focus on convenience but also to address the associated risks and side effects adequately. Cyber risks are omnipresent, meaning that any business can become a target, and cyberattack on critical infrastructure can have unforeseeable, far-reaching, and severe consequences.

Within the maritime sector, safety has always been paramount, and this has obviously been strongly embedded in education as well. Yet, this hierarchical industry, which has historically been slow to adapt to digitalisation, has become a prime target in cyberspace in recent years. The rapid rise of digital adoption and internet connectivity in this sector (IT—information technology, IoT—Internet of things, OT—operational technology, and IoE—Internet of Everything, automations, autonomous vessels, etc.) has not been matched by a corresponding emphasis on cybersecurity or even by enhanced digital proficiency. The cyberattack on Maersk served as a stark reminder of the potential damage a single attack can cause. This article will focus on how to prepare the people working in this domain for this highly vulnerable, yet critical sector, where up to 90% of the international trade takes place (according to the International Maritime Organisation—IMO) to meet the new digital challenges while considering cybersecurity [1, 2].

It is not enough to rely on the assumption that the next generation will inherently understand these challenges better. Digital natives are not born with an innate understanding of cybersecurity, especially in the context of maritime; they are simply confident and bold users of technology, with all the benefits and drawbacks of this trait. Therefore, education and practical training will remain crucial. Moreover, there is a pressing need for regulatory guidance from organisations like the IMO and international regulatory bodies, as the cost of implementing cybersecurity measures will be avoided until it is enforced. Efforts as such can drive the goal of ensuring that the industry has a unified global resilience, recovery, and response capabilities ensuring safety, security, and business continuity.

2 Threats, Incidents, and Counter Actions

Maritime cybersecurity has become a very important issue, due to the several cyberattacks hitting the sector and resulting in significant financial losses, critical assets as well as lowering reputation-based trust for the affected organisations. Cyber incidents involving navigation, movement of cargo, and other processes threaten lives, the environment, property, and considerably disrupt maritime trade movement. Compared to other areas of security, managing cyber risk in the maritime domain is more challenging due to the lack of information about the nature and impacts of these attacks [3].

Since the start of the COVID-19 pandemic, cyberattacks on maritime systems have surged by 400% between 2020 February and June, while ports and harbours have seen a 900% increase in attacks between 2017 and 2021, with threats such as ransomware, phishing, and malware targeting both vessels and port infrastructure. For instance, the Port of Houston experienced a breach due to vulnerabilities in password management systems that allowed attackers to gain network access [4–6].

Incidents naturally prompted the sector to take countermeasures and regulatory bodies like the IMO have established guidelines, the EU has revised Network and Information Systems (NIS) directive, and NIS2 is also soon in force to help counter these risks and emphasise the importance of robust cybersecurity measures. Despite these efforts, cyber incidents are still on the rise in 2024, largely due to human error and in many cases non-compliance. Although the number of attempted attacks do not incline so steep any more, the ransom payouts by maritime organisations have significantly increased [3].

Stakeholders of the maritime cybersecurity domain view cybersecurity problems through different lenses. This article will approach maritime cybersecurity from a civilian point of view that lies in the cross-section of the full spectrum of the maritime cyber domain: operations and management, economic and financial aspects, data challenges, policies and regulations, and training and education of the active workforce and decision makers. Whichever aspect is examined more closely, the role of human factor cannot be overlooked.

2.1 *Economic and Financial Aspects*

Economic studies highlight the severe financial impact of cyberattacks on the maritime industry. Maritime sector is still less prepared compared to other sectors like banking or aviation. Cybersecurity breaches are often underreported, leading to a lack of comprehensive data on the financial losses. A proactive approach, including scenario-based planning and updated insurance practices, is necessary to address uncertainties and aid stakeholders to more conscious operations. The World Economic Forum reported substantial financial losses due to cyberattacks across sectors, including maritime. The NotPetya cyberattack on Maersk in 2017 alone resulted in losses of approximately USD 300 million. Furthermore, major ports globally have faced an average of 12 attacks per day, and ship owners typically pay about USD 3.1 million in ransom per attack. The overall economic impact extends beyond immediate financial losses to include operational disruptions, compromised safety, environmental damage, and reputational harm. The 2021 blockage of the Suez Canal, even though not related to a cyberattack, cost as much as 12% of global trade, meaning that should a cyberattack on such a ship at the Suez Canal occur, a similar loss would be incurred [4, 7].

Ransom demands have increased by more than 350% in 2022 compared to previous year. The total costs of cyberattacks can be attributed to ransom payments, system recovery, and business interruption, underscoring the growing financial burden on the maritime sector [8].

In 2024, it is no longer true that shipping companies are simply not investing in cyber risk management. Progress is being made, even if this is still in the early stages of maturity. To reveal deeper insights, an analysis has been performed across shipping companies worldwide to get a better understanding of where defensive measures are being strengthened, and which areas still require significant work. It was found that beyond the obvious costs of implementing cybersecurity measures—like antivirus software, consulting, network security, and training—hidden costs often go unnoticed. These can include additional satcom bandwidth, extra cloud storage, and the expense of added human resources needed to manage cybersecurity. Such factors significantly contribute to the overall cost of cyber protection but are often overlooked [8].

The main costs of cybercrime come from business interruptions, delays, and the expense of restoring or replacing systems. However, the impact on a company's reputation can be even more damaging, with long-lasting effects that can be slow or even impossible to recover from.

2.2 Risk Appetite

Experts argue that the costs of consequences can be minimised by dedicating the right resources to security operations. However, within the maritime domain, matching the necessary cyber experts with those who understand this niche industry is a very present challenge. Many ship owners are used to tolerating high risk in general, and it seems this applies to cyber risk as well. For some, the approach is simply to meet the standards at minimum requirement.

As costs associated with cyberattacks are on the rise, shipping operators too are turning to insurance; however, uncertainty around cyber risks complicates policies and claims. Research shows that 72% of companies lack cyber insurance, and 92% of maritime cyberattack costs go uninsured. Incident resolution takes on average 57 days, slightly longer compared to non-maritime domains, due to the need for coordination between offshore experts and onboard crews. Behind these statistics, the probable reason is that ship owners do not meet insurance requirements to be eligible for insurance, or prefer to accept high risks, opting to meet only the minimum standards. The challenge with this cost-based approach is that it focuses on avoidance or at least shifting the financial responsibility to mitigate the cyber incident's cost, but does not invest in human cyber capacity building that could enable more resilient and sustainable operations, while, with poor visibility and slow responses increasing the risk of incidents escalating and leading to greater losses [5, 9].

Quantifying the total cost of cyber risk remains challenging, making it hard to secure resources for effective management, since it focuses on reducing risks rather than boosting revenue or cutting costs. UK government research shows that boards often struggle with cybersecurity due to gaps in knowledge, training, and time, underscoring the need for cyber professionals to make strong business cases, particularly to finance leaders [10]. The result of a 2023 survey illustrate the challenge in shipping: 33% felt that one of the biggest challenges in improving cyber risk management is understanding the level of risk and 30% said that it was difficult to understand best practice [11].

Smaller operators may manage ships as complex as industry leaders, facing equal risks of cyberattacks. However, with fewer resources, they must rely on external help or accept higher risk tolerance compared to larger counterparts. Investing in inadequate or inappropriate resources can negatively impact an organisation in less obvious ways. Ship operators often focus on the initial costs of cybersecurity solutions but overlook the on-going resources needed to maintain these systems. For instance, setting up individual logins and passwords seems like a sound security measure, but in practice, it incurs on-going costs to keep credentials updated amid changing crews and visitors illustrating how maritime is different from other sectors. There is already an obvious need for combined maritime and cyber skills, but attracting cyber talent into shipping is a major issue at present [12].

2.3 Human Element

There is no technical solution for bad culture. A lack of talent or human failure will be responsible for over 50% of all significant cyber incidents by 2025, according to Gartner [13]. The shipping industry needs a major refocus on people to minimise the chance of this predicted statistic becoming a reality. Challenge is present for several reasons. Firstly, cyber professionals are expensive, and with their booming job market, which has less than 1% unemployment, maritime companies would need to offer high salaries to attract them. Secondly, maritime cybersecurity differs from traditional IT, often involving outdated systems and requiring flexible, practical solutions that may clash with standard cybersecurity controls, making the field less attractive to specialists. Lastly, cybersecurity roles in shipping are usually part of IT and often involve travelling to remote and harsh locations, which does not appeal to professionals accustomed to working remotely.

Cybersecurity is a horizontal function that connects all tasks; this way ensuring the safety and integrity of every operation, both on- and offshore. Rather than treating it as a separate responsibility, cybersecurity skills and capabilities should be embedded into every process, making it a core element of daily operations, decision-making, and risk management within the maritime workforce. This should entail extended training curriculum.

2.4 Training, Education, and Curriculum Development

Poorly trained staff in all maritime-related organisations increases the risks of cybersecurity of the sector. Studies conducted in the past 5 years reveal a critical need for proper employee training within maritime institutions, as well as need for implementing organisational frameworks. Even though plans exist, failing to implement them makes organisations ever more vulnerable to cyberattacks. Training is viewed as a proactive defence measure that enhances overall resilience. Researchers emphasise the need for updating training programmes to equip marine cadets, future seafarers, and professionals with relevant skills. In this regard, Larsen and Lund (2021) conducted a review of how the human dimension perceives cyber risk. They employed psychological models to accomplish their goal. The review showed that there is currently a lack of studies on the human behaviour towards cyber risk within the maritime industry. Designing effective training programmes to adequately equip maritime professionals is only possible with a thorough understanding of the psyche of the participants [14].

Already in 2015, Caesar and Choon highlighted the need for Maritime Education Training (MET) institutes to sample the psychological profiles of cadets when admitting them to tailor training packages that meet the expectations of all stakeholders. They argued that there is a need for a conscious departure from the orthodoxy that characterises the training of maritime industry experts to provide a more holistic and future-ready skill set that strategically positions them to deal with the growing complexity of maritime cybersecurity threats [15].

A recent study from 2021 already demonstrated how training could potentially evolve for maritime cyberspace. Using the development of an interactive environment for educational purposes, simulated environment was created for training with cyber threats mainly for vulnerability analysis and other maritime cybersecurity exposures, like traffic eavesdropping and GNSS/AIS spoofing. At present, only very few such environments are available worldwide. In terms of training, there is a need for renewal. It is not enough to only instruct, having the right people with the right skillset may require different approach in preselection of the future seafarers, to be able to deal with maritime cybersecurity threats. Some literature on training conclude that the vulnerabilities lie in the human element of maritime system management, and state that many people working in the maritime sector today have limited knowledge of cybersecurity issues. This indicates the need for increased training to improve awareness levels as a form of defence against future cyberattacks [4].

2.5 Operations and Management

There are numerous studies on the operations and management aspects of cybersecurity. Generally, maritime organisations tend to be weak in their cybersecurity activities, poorly integrating organisational structures against cyber threats. There are systems and applications created to perform threat identification, but system design should start from a systems approach. Maritime systems are highly complex and include legacy elements, where any unvetted third-party partner can increase risks in the supply chain. For example, some major problems may arise when individual company systems are not properly maintained, thereby exposing numerous vulnerabilites which expose them to attacks. To counter these concerns, interface training, hardware replacement, and the introduction of security measures must be implemented, along with a collaborative approach between experts and end users of maritime cyber systems. Studies show that most companies have not yet conducted a cybersecurity audit to identify how their systems perform in the face of new threats and may not do so in the foreseeable future unless prompted by an actual attack [4]. This indicates the 'opting for the minimum' approach of many ship owner companies.

2.6 Policies and Regulations

Numerous existing maritime cybersecurity policies and regulations are outdated and often inadequate. Studies suggest the need for comprehensive international frameworks, stricter enforcement, and better communication of threats. There is a call for updated regulations to address emerging challenges, such as Internet of Things (IoT) devices and autonomous ships, and for collaboration among global stakeholders to strengthen cybersecurity laws. Policymakers' and regulators' roles are important in providing incentives to motivate corporations towards engaging in activities and adoption of relevant measures to lessen the chances of deliberate cyberattacks on supply chains [4].

This certainly will be a long process which will take place gradually. Existing policies include the IMO Resolution MSC.428(98) implemented in 2021 that requires owners, operators, and managers to consider overall cyber risks, and to implement cybersecurity across all levels of their management system, in line with International Safety Management (ISM) Code.

International Association of Classification Societies' (IACS) new unified requirements (URs) for cybersecurity in effect from 1 July 2024 oblige owners, yards, and suppliers to build cybersecurity barriers into their systems and vessels, requiring compliance across the full spectrum of critical on-board control and navigation systems. Its two sets of rules involved (IACS UR E26. governs system integration, while UR E27 applies to essential on-board systems) are technical regulations aimed at ensuring the security of computer-based systems on board ships. They mention human oversight in implementation and compliance, system integration, operation and management, and third-party verification. However, it does not specifically mention or encourage training on the human element.

The EU NIS2 Directive implemented in October 2024 obliges EU member states to adopt cybersecurity strategies and establish competent cybersecurity structures across their jurisdictions. Under NIS2, the maritime sectors, specifically entities such as ports, shipping companies, and other maritime transport operators, are recognised as essential services. These entities are required to implement appropriate security measures and report significant cybersecurity incidents. The goal is to strengthen the overall cybersecurity resilience of critical infrastructure, including the maritime industry, which is integral to global trade and economic stability. The NIS2 already advocates security and awareness trainings, incident response exercises, and cybersecurity skills development in general. Thus, maritime organisations operating within the EU must comply with NIS2 requirements, including implementing robust cybersecurity measures and reporting incidents, or they may face penalties and sanctions for non-compliance.

Similar legislation is expected to take effect in other key regions over the coming years [16].

2.7 Data Issues

Data quality is a critical issue in maritime cybersecurity, affecting decision-making and risk assessments—just take the example of how inaccurate Automatic Identification System (AIS) data can negatively affect the decision-making of the shipping crew. Data is being created every minute. Ship sensors provide real-time information on vessel performance and condition, from weather reports through sailing routes, to port information on cargo data, berthing schedules, and more. The data owner will have to decide how to handle, store, safeguard, and use this continuously generated data. However, reluctance to share sensitive information hinders research, making qualitative approaches to be less precise. It is important to note that consequences of data breaches could be severe; therefore, better data management and independent assessments are essential to maintain and strengthen cybersecurity in the field.

3 The Way Ahead

The maritime sector's escalating dependence on digital technologies, coupled with the increasing complexity of cyber threats, necessitates a proactive and multi-dimensional approach to address the cybersecurity skills gap. To ensure the long-term security and resilience of maritime operations, the following strategic initiatives are recommended.

3.1 Development of Continuous Cybersecurity Training Programmes

The initial and most essential step in addressing the cybersecurity skills gap within the maritime domain is the development of continuous and structured training programmes specifically tailored to the sector's unique needs. Traditional, one-time training sessions are inadequate in the context of rapidly evolving cyber threats, and they do not give the 'sets and reps' but provide an ad-hoc experience. Instead, there is a critical need for a dynamic training ecosystem that evolves in parallel with the threat landscape and ensures continuous engagement with practical outcomes.

These training programmes should be designed to address the needs of all workforce levels, from entry-level seafarers to senior shore-based personnel. A tiered training approach ensures that each group receives instruction that is relevant to their specific roles, whether it involves fundamental cybersecurity practices, such as safe password management or more advanced threat detection and response strategies. Moreover, these programmes should incorporate practical, scenario-based exercises that replicate real-world cyber incidents, just like the maritime industry is conducting its safety drills.

3.2 Integration of Cybersecurity into Maritime Education and Training

While continuous training programmes; and drills are essential for the current workforce, a long-term solution requires embedding cybersecurity education within the foundational training provided by maritime academies and higher education institutions (HEIs). The current curricula in many of these institutions do not sufficiently address the specific cybersecurity challenges that maritime professionals will encounter in their careers, partially for the lack of cybersecurity competence and experience among the maritime domain professors and educators. This gap must be bridged by revising and expanding educational programmes to include comprehensive cybersecurity modules, and by educating HEI educators on the topic as well.

These modules should cover a wide range of topics, from basic concepts like understanding cyber threats and vulnerabilities to more advanced subjects such as incident response and the protection of critical maritime infrastructure. The curriculum should be developed in close collaboration with industry experts to ensure that it remains relevant and aligned with the latest industry standards. Moreover, the curriculum must be flexible enough to adapt to the fast-changing nature of cyber threats, with regular updates that reflect the latest developments in both technology and threat intelligence. As the digital transformation continues such modules will serve as a core foundational layer to the emerging technologies like always-on-connectivity, automation, and autonomous decision-making in the maritime sector.

3.3 Strengthening Industry-Expert Collaboration

Collaboration between the maritime industry and cybersecurity experts is essential for developing training programmes and educational content that are both relevant and effective. Such partnerships can help bridge the gap between theoretical knowledge and practical application, ensuring that maritime professionals are well-prepared to handle the specific cyber threats they are likely to face. One proposed initiative is the establishment of cybersecurity mentorship and exchange programmes. Through these programmes, cybersecurity experts could work directly with maritime personnel, providing hands-on training and sharing best practices. This close collaboration would allow for the customisation of cybersecurity strategies to meet the unique needs of different maritime organisations, whether they operate large fleets or manage critical port infrastructure.

3.4 Implementation of Regulatory and Incentive-Based Approaches

The role of governments and international regulatory bodies, such as the IMO, is crucial in setting global standards for maritime cybersecurity. These entities should take the lead in developing and enforcing comprehensive cybersecurity frameworks that ensure a consistent level of security across the industry. Regulatory standards should be designed to address the full spectrum of cyber risks, from protecting individual vessels to securing the entire supply chain, but allowing the flexibility of addressing future challenges.

In addition to setting standards, governments and industry leaders should provide financial incentives to encourage the adoption of robust cybersecurity practices. For example, tax breaks or grants could be offered to companies that invest in cybersecurity training programmes and infrastructure. These incentives would lower the financial barriers to implementing necessary cybersecurity measures, making it easier for smaller operators to comply with new regulations and improve their overall security posture. Another incentive is penalising non-compliance, such as the mechanism established by the NIS2 directive in the EU, where organisations that fail to meet mandated cybersecurity standards can face substantial fines and other regulatory sanctions.

Moreover, regulatory bodies should consider introducing mandatory cybersecurity audits and certifications for maritime organisations. These audits would assess an organisation's adherence to established cybersecurity standards and identify areas where improvements are needed. Certification could then be used as a benchmark of cybersecurity maturity, with certified organisations being eligible for certain benefits, such as reduced insurance premiums or preferential treatment in contracts.

3.5 Leveraging the Insurance Industry

The insurance industry can play an important role in promoting better cybersecurity practices within the maritime sector. Insurance companies have the unique ability to influence organisational behaviour by linking insurance premiums to cybersecurity performance. Companies that demonstrate strong cybersecurity measures and a history of low cyber incident rates could be rewarded with lower insurance premiums, creating a financial incentive for continuous improvement. To support this, the development of a cybersecurity certification process for maritime organisations could be implemented (as mentioned in the paragraph above). Additionally, insurance companies could offer tailored cybersecurity insurance products that address the specific risks faced by maritime operators, further aligning financial incentives with improved cybersecurity outcomes.

3.6 Encouraging a Culture of Continuous Improvement

Finally, fostering a culture of continuous improvement is essential for maintaining a high level of cybersecurity readiness in the maritime sector. Just as safety drills are a regular part of maritime operations, cybersecurity drills should be conducted routinely across all levels of an organisation. These drills would help ensure that all personnel are familiar with their roles in responding to cyber incidents and that the organisation is prepared for the eventuality of an attack. To support this culture, maritime organisations should also encourage on-going professional development in cybersecurity. This could include providing access to advanced courses, certifications, and workshops that keep maritime professionals up to date with the latest developments in the field. By embedding cybersecurity into the daily operations and decision-making processes, organisations can ensure that it becomes a core component of their overall risk management strategy.

4 Conclusion

The maritime sector is at a critical point where prioritising cybersecurity skills development is crucial for protecting its infrastructure against increasingly complex cyber threats. While innovative training methods and addressing current challenges can significantly boost the sector's resilience, the role of maritime organisations and stakeholders in enforcing regulations and providing top-down guidance remains essential. Key organisations like the IMO, classification societies, and national maritime authorities are instrumental in setting standards, providing guidance on best practices, issuing directives, and ensuring compliance across the industry. However, despite all efforts, gaps in cybersecurity readiness are expected to persist, ranging from inconsistent adoption of standardised training protocols, regional variations in regulatory enforcements, and the rapid evolution of cyber threats that outpace security measures. The integration of new technologies like artificial intelligence and automation may introduce unforeseen vulnerabilities that the workforce is not yet fully equipped to handle. To bridge gaps, future research should aim to develop standardised, globally recognised cybersecurity training protocols and continuously evaluating the effectiveness of these to adapt to the changing threat landscape. Regulatory bodies will provide a necessary framework, but the responsibility for sound cybersecurity decisions ultimately lies with the workforce operating within this critical infrastructure, it will never be only technology. Achieving a cyber-resilient sector requires a collective effort from all stakeholders, with the adoption of regular cybersecurity exercises alongside the traditional safety drills. Comprehensive training with updated operational practices, regulatory reforms, and improved data management will all be vital components for enhancing maritime cybersecurity.

Acknowledgments Research for this publication was funded by the EU Horizon2020 project 952360-MariCybERA.

References

1. J.M. Lane, M. Pretes, Maritime dependency and economic prosperity: why access to oceanic trade matters. Mar. Policy **121**, 104180 (2020). https://doi.org/10.1016/j.marpol.2020.104180
2. These are the world's most vital waterways for global trade, World Economic Forum, https://www.weforum.org/agenda/2024/02/worlds-busiest-ocean-shipping-routes-trade/. Accessed 14 Apr 2024
3. N. Kala, M. Balakrishnan, Cyber preparedness in maritime industry. Int. J. Sci. Tech. Adv. **5**(2), 19–28 (2019)
4. M. Afenyo, L.D. Caesar, Maritime cybersecurity threats: gaps and directions for future research. Ocean Coast. Manag. **236**, 106493 (2023)
5. Maritime industry sees 400% increase in attempted cyberattacks since February 2020, Security Mag. (2020), https://www.securitymagazine.com/articles/92541-maritime-industry-sees-400-increase-in-attempted-cyberattacks-since-february-2020. Accessed 31 Aug 2024
6. Axio, *Port of Houston Prevents Data Breach: A Success Story Highlighting the Importance of Privileged Access Management (PAM) Controls* (Axio), https://axio.com/insights/port-of-houston-data-breach/. Accessed 31 Aug 2024
7. Top Cybersecurity Statistics for 2024 | Cobalt, https://www.cobalt.io/blog/cybersecurity-statistics-2024. Accessed 31 Aug 2024
8. CyberOwl, *Maritime Cyber Risk Report: Shipping Industry Remains "Easy Target", Pays Average US$3.2m in Cyberattacks* (CyberOwl), https://cyberowl.io/maritime-cyber-risk-report-shipping-industry-remains-easy-target-pays-average-us3-2m-in-cyberattacks/. Accessed 6 Aug 2024
9. Meeting the cyber threat challenge in the maritime industry—protection beyond regulation, Maritime London, https://www.maritimelondon.com/news/meeting-the-cyber-threat-challenge-in-the-maritime-industry-protection-beyond-regulation. Accessed 28 Aug 2024
10. Cyber security breaches survey 2023, GOV.UK, https://www.gov.uk/government/statistics/cyber-security-breaches-survey-2023/cyber-security-breaches-survey-2023. Accessed 28 Aug 2024

11. N. Chubb, *How Can Cyber Risks Be Managed in a Technologically-Evolving Maritime Industry?* (Thetius), https://thetius.com/how-can-cyber-risks-be-managed-in-a-technologically-evolving-maritime-industry/. Accessed 29 Aug 2024
12. WEF_Global_Cybersecurity_Outlook_2024.pdf, https://www3.weforum.org/docs/WEF_Global_Cybersecurity_Outlook_2024.pdf. Accessed 29 Aug 2024
13. Cybercrime and its growing threat in financial services, 6point6, https://6point6.co.uk/insights/growing-threat-of-cybercrime-in-financial-services/. Accessed 28 Aug 2024
14. M.H. Larsen, M.S. Lund, F.B. Bjørneseth, A model of factors influencing deck officers' cyber risk perception in offshore operations. Marit. Transp. Res. **3**, 100065 (2022). https://doi.org/10.1016/j.martra.2022.100065
15. L. Caesar, S. Cahoon, Training seafarers for tomorrow: the need for a paradigm shift in admission policies. Univers. J. Manag. **3**(4), 160–167 (2015). https://doi.org/10.13189/ujm.2015.030404
16. Preparing for IMO's ISM cyber security, DNV, https://www.dnv.com/maritime/insights/topics/maritime-cyber-security/regulations/. Accessed 30 Aug 2024

Open Access This chapter is licensed under the terms of the Creative Commons Attribution-NonCommercial-NoDerivatives 4.0 International License (http://creativecommons.org/licenses/by-nc-nd/4.0/), which permits any noncommercial use, sharing, distribution and reproduction in any medium or format, as long as you give appropriate credit to the original author(s) and the source, provide a link to the Creative Commons license and indicate if you modified the licensed material. You do not have permission under this license to share adapted material derived from this chapter or parts of it.

The images or other third party material in this chapter are included in the chapter's Creative Commons license, unless indicated otherwise in a credit line to the material. If material is not included in the chapter's Creative Commons license and your intended use is not permitted by statutory regulation or exceeds the permitted use, you will need to obtain permission directly from the copyright holder.

Rethinking Seafarer Training for the Digital Age

Dan Heering

1 Introduction

The maritime industry plays a vital role in global trade, with over 80% of the world's goods transported by sea [1]. As of January 2023, the global maritime fleet includes 105,493 vessels with a combined capacity of 2.27 billion deadweight tons (dwt), underscoring its vast scale and essential role in sustaining the global economy [2]. By 2030, the global maritime digital technology industry is expected to be worth $345 billion, up from a previous estimate of $279 billion [3].

In recent years, the maritime industry has made significant advancements including the increased use of digital systems, improvements in satellite communications, the development of IoT-enabled port infrastructure, and trends towards autonomous shipping [4–7]. While much attention has been placed on technological progress, a critical area that remains under-addressed is cybersecurity. The increasing reliance on interconnected technologies, such as IoT networks, automated systems, and advanced telecommunications, has significantly increased the maritime sector's exposure to possible cyber threats. As ships, ports, and other maritime infrastructures become more digitised, they present numerous vulnerabilities that can be exploited by malicious actors. While these technologies have greatly enhanced efficiency and operational capacity, they have also introduced new vulnerabilities, making the industry more susceptible to cyberattacks such as ransomware, phishing schemes, and system breaches.

For example, the 2017 NotPetya attack, which disrupted A.P. Møller-Maersk's operations, highlighted the significant impact cyberattacks can have on the shipping

---

D. Heering (✉)
Estonian Maritime Academy, Tallinn University of Technology, Tallinn, Estonia
e-mail: dan.heering@taltech.ee

industry, endangering the safety of vessels, crew, and cargo [8]. Several other cyberattacks demonstrate the vulnerability of maritime systems:

- In July 2018, COSCO Shipping Lines experienced a ransomware attack that affected its operations across the Americas. While the company was able to isolate the attack, it caused significant delays and disrupted communication systems between COSCO's U.S. offices and its worldwide shipping network [9].
- The Port of San Diego and the Port of Barcelona suffered ransomware attacks in 2018 severely disrupting their IT systems. The attacks not only delayed port operations but also raised concerns about the broader vulnerability of port infrastructure to cyberattacks [10].
- Similarly, in February 2020, the International Maritime Organization (IMO) itself was targeted by a cyberattack that took its public website and internal systems offline. Although the attack did not compromise core operational systems, it served as a wake-up call regarding the cyber resilience of international maritime institutions [11].

Naval Dome, an Israeli cybersecurity firm specialising in maritime systems, conducted in 2017 a series of cyber penetration tests on critical systems aboard various ships, including tankers, container ships, and cruise ships. These tests, performed with permission from system manufacturers and owners, revealed how easily hackers could access and override key systems such as navigation, radar, and machinery controls. One test involved altering a vessel's reported position via email, misleading the crew into thinking the ship was on course, when in reality, it could have run aground. Another attack disrupted machinery operations, tampered with fuel and ballast systems, and erased radar targets, all while the ship's systems continued to show normal functionality on their displays [12]. These tests exposed serious vulnerabilities, showing that even sophisticated maritime security systems could be compromised with relative ease, often through routine operations like system updates.

These incidents highlight the growing sophistication of cyber threats in the maritime industry and emphasise the need for robust cybersecurity measures onboard ships and across maritime infrastructures. Cyber threats are not limited to targeting corporate offices but can affect ships at sea, making it crucial that seafarers themselves are equipped to manage these risks.

Despite these risks, cybersecurity awareness among seafarers remains alarmingly low. Many seafarers are still unaware of the potential cyber risks they face, both in their personal lives and their professional roles onboard ships. Recent studies reveal that a large portion of seafarers have not received any form of cybersecurity training, despite the growing threats they face while at sea [13–16]. This lack of awareness is further compounded by a weak cybersecurity culture within many shipping companies, where cybersecurity practices are often seen as secondary to operational concerns [17, 18]. Without a strong foundation of cybersecurity practices, seafarers may unknowingly expose critical systems to attack through simple mistakes, such as plugging an unsecure USB into a navigation system or falling victim to phishing schemes.

The situation is further complicated by the easy access to the internet that seafarers now can have onboard the ships. The Maritime Labour Convention (MLC), which advocates for better living and working conditions for seafarers, has made it possible for crew members to have greater access to internet services while at sea [19, 20]. While this has been a significant improvement for the welfare of seafarers, allowing them to stay in touch with family and friends, it also widens the possibility of future cyber incidents onboard ships. Many seafarers use their personal devices to connect to the ship's network, which, if not secured properly, can lead to the introduction of malware or other security breaches. The recent petition for free internet access for seafarers has further fuelled the debate, with some arguing that greater internet access could lead to more exposure to cyber risks, while others emphasise the importance of connectivity for crew morale [21].

This weak focus on cybersecurity creates gaps in awareness and readiness, leaving the maritime industry highly exposed to advanced cyberattacks. Without strong leadership and prioritisation of cyber resilience, the sector is at greater risk of facing more disruptions.

The International Maritime Organization (IMO) has responded to the rise in cyber threats with guidelines such as Resolution MSC.428(98) on Maritime Cyber Risk Management, urging companies to include cyber risks in their Safety Management Systems (SMS) [22]. However, the practical implementation of these guidelines is uneven across the industry. While some companies have begun to adopt more robust cyber risk management practices, many others lag behind, leaving their operations exposed. Furthermore, the International Convention on Standards of Training, Certification, and Watchkeeping for Seafarers (STCW) still lacks comprehensive provisions for cybersecurity awareness, making it clear that cybersecurity training for seafarers is not yet seen as essential [13]. While the STCW convention has provided the framework for essential maritime training, its current provisions lack the comprehensive cybersecurity guidelines needed for today's digital maritime environment. For true cyber resilience, the STCW convention will need to adapt and introduce compulsory cybersecurity training that covers both technical skills and behaviour-focused models.

To address these challenges, this chapter proposes a novel approach to cybersecurity education for seafarers. By focusing on behaviour change models, we aim to foster a more proactive attitude towards cybersecurity, encouraging seafarers to adopt better practices both at sea and in their personal lives. Rather than relying only on theoretical knowledge, the proposed training framework seeks to transform how seafarers respond to cyber threats, developing practical habits that promote cyber resilience.

A detailed case study from TalTech Estonian Maritime Academy demonstrates how compulsory cybersecurity training can foster a security-conscious culture among future maritime officers. The case study highlights the importance of integrating behaviour-focused training into maritime curricula, showing how future seafarers can develop the skills needed to navigate the digital risks of modern maritime operations. This chapter offers practical insights and strategies for reshaping maritime cybersecurity education to meet the challenges of the digital age.

2 Cybersecurity Challenges in Maritime Operations

Until recently, cybersecurity in the maritime domain was not seen as a significant issue. However, as the industry has become increasingly reliant on digital technologies such as IoT (Internet of Things), automation, and advanced communication systems, cyber threats have grown exponentially. These advancements, while improving operational efficiency, have also introduced new vulnerabilities that can be exploited by cybercriminals. The interconnected nature of modern maritime systems, from ships to ports, presents an extensive attack surface for malicious actors.

The European Union Agency for Cybersecurity (ENISA) carried out a study on cybersecurity challenges in the maritime industry in 2011. The study aimed to help the sector better understand its key cybersecurity risks, targeting organisations, national authorities, government bodies, and private companies involved in maritime activities [23]. The report provided several key findings:

- Low cybersecurity awareness: Awareness of cybersecurity risks is either at a very low level or non-existent across the maritime sector. This applies at all levels, including government bodies, port authorities, and maritime companies.
- Complex ICT systems: Information and Communication Technology (ICT) systems supporting maritime operations, from port management to ship communication, are highly complex and incorporate various specialised technologies. The rapid pace of technological development, coupled with the trend towards increased automation, has led to a reduced focus on security features in many cases. It was noted that several maritime governance stakeholders relevant to EU Member States exist. However, there is an apparent lack of coordination between these organisations regarding cybersecurity and the risks associated with cyber threats.
- In the current regulatory context (in 2011) for the maritime sector on global, regional, and national levels, there is very little consideration given to cybersecurity. Most security-related regulation only includes provisions relating to safety and physical security concepts. These are found in the International Ship and Port Facility Security (ISPS) Code and other relevant maritime security and safety regulations. Regulations such as (EC) No 725/2004 on enhancing ship and port facility security do not consider cyberattacks as possible threats of unlawful acts.
- No holistic approach to maritime cyber risks exists. It was observed that maritime stakeholders are setting and managing cybersecurity expectations and measures in an ad hoc manner. Not all risks are being considered, such as the disruption of critical telecommunication means or the exposure of cargo information.
- The key stakeholders of the maritime sector still lack the necessary incentives to improve their overall cybersecurity posture. This results from a combination of fragmented and insufficient regulatory frameworks that do not adequately address security aspects. These range from a lack of good security baselines to a poor range of direct economic incentives to implement good security.

- Despite the lack of a holistic and comprehensive approach towards cybersecurity, several initiatives are being implemented that may contribute to national Critical Information Infrastructure Protection (CIIP) efforts.

One of the high-level recommendations made in the ENISA study was the need for the development and implementation of awareness-raising campaigns. These should specifically target the maritime sector and include cybersecurity training for relevant stakeholders, such as shipping companies, ship crews, and port authorities.

The Maritime Unmanned Navigation through Intelligence in Networks (MUNIN) project explored between 2012 and 2015 the feasibility of unmanned bulk carriers during intercontinental voyages. While the study identified and classified 65 hazards based on their consequences and likelihood of occurrence, risks related to the cyber domain were not included [24].

Fitton et al. address cyber operations in the maritime domain by focusing on three key elements: information, which involves safeguarding data and communication systems; technology, which requires securing hardware, software, and networks; and people, who must be trained to recognise and respond to cyber threats. Ship crews now require constant internet access to relieve the challenges of long periods at sea, but this increased connectivity also makes them more vulnerable to cyberattacks. The authors emphasise the need for education, training, and regular drills to prevent, identify, and defend against cyberattacks, ensuring that operations can continue even under attack. They also highlight the importance of understanding social engineering tactics and recommend specialised training to mitigate these risks [25].

Di Renzo et al. explore the vulnerabilities present in shipboard systems, oil rigs, cargo handling, and port operations. They emphasise that shipowners must ensure their personnel are adequately trained and equipped to both prevent and swiftly recover from cyberattacks. This preparation should be an integral part of a comprehensive risk management strategy that identifies potential threats, vulnerabilities, and their consequences. The study highlights key issues such as insufficient awareness at the management level, a lack of detailed information on cyberattacks and vulnerabilities, and inadequate cybersecurity training for personnel [26].

The significance of cybersecurity training for shipping companies and personnel began to receive more substantial attention from researchers around 2016–2017 [27–35]. A pivotal moment came with the A.P. Møller-Maersk cyber incident in June 2017, which highlighted the severe vulnerabilities of shipping companies and port infrastructures to technological failure [8]. This incident acted as a wake-up call for the maritime industry, prompting a surge in academic and industry publications focused on identifying and addressing cyber threats and vulnerabilities in shipping operations. Following 2017, there was a noticeable increase in research and initiatives aimed at strengthening maritime cybersecurity practices.

Articles published between 2018 and 2020 offered preliminary recommendations for improving maritime cybersecurity training, focusing on addressing the industry's growing vulnerability to cyber threats [36–42]. These studies identified several critical gaps in cybersecurity awareness among maritime professionals and emphasised the need for practical, scenario-based training, as well as the incorporation of cybersecurity into maritime education curricula. Early recommendations highlighted the importance of training ship officers to recognise and respond to cyber threats, particularly phishing attacks and malware, while also fostering a culture of cyber hygiene onboard vessels.

The vigilant seafarer onboard is arguably the most critical security asset for any shipping company, as they are directly responsible for monitoring and responding to potential cyber threats. Hareide et al. emphasise the need for a high degree of cyber situational awareness, particularly for navigators [38]. To make well-informed decisions, navigators must maintain a comprehensive understanding of their IT systems, including their capabilities and limitations. Without this knowledge, operators may become a risk factor, introducing vulnerabilities rather than mitigating them.

Building the human capacity to serve as the strongest link in the maritime cybersecurity chain requires a focus on education and training. Lund et al. highlight the importance of enhancing a navigator's competence through increased system awareness, which is essential for reducing cyber risks onboard. This involves not only technical training but also fostering a culture of vigilance and proactive defence measures to prevent and respond to cyber threats effectively [43].

Roolaid hypothesised in his dissertation that maritime educational institutions have not placed sufficient emphasis on specialised cybersecurity education in the training of deck officers, potentially compromising their ability to operate ships safely in the face of growing cyber threats [44]. To explore this, the author conducted a series of surveys and interviews with European maritime educational institutions to assess their approach to cybersecurity training for seafarers. The study focused on analysing the curricula, courses, and learning outcomes related to cybersecurity across these institutions. The research gathered information from approximately 35 maritime educational and training (MET) institutions, with the following key findings:

- Only 2 out of 35 institutions provided specific cybersecurity education tailored for ships' officers.
- Three institutions offered general cybersecurity awareness education for ships' officers.
- Of the 19 institutions that responded fully to the survey, 11 considered it necessary to teach cybersecurity to ships' officers.
- About 7 out of 19 institutions believed that cybersecurity education would be essential in the future.
- One institution did not see the need for cybersecurity education for ships' officers.

These findings suggest a significant gap in the cybersecurity preparedness of maritime professionals. The limited focus on specific cybersecurity training for ship officers indicates a pressing need for maritime educational institutions to adapt their curricula to meet the challenges of a digitised maritime industry. The results of the survey concur with the reports and industry guidelines that recommend educating and training of students and active seafarers. However, the collaboration between shipping companies and educational institutions in Europe is still lacking the cybersecurity component.

Roolaid's research identifies several significant obstacles to provide effective cybersecurity training for seafarers. These include the excessive workloads already borne by educators, a lack of available study materials in native languages, an overburdened curriculum, and the absence of specific requirements in the STCW Code to mandate such training. Similarly, [39] also highlight these challenges, pointing out that the current structure of maritime education does not adequately support cybersecurity readiness.

To address these gaps, Roolaid recommends implementing a concise, 2-day cybersecurity course specifically for ship officers. This course would provide a combination of theoretical knowledge on maritime cybersecurity, focusing on key concepts such as threat identification and risk assessment, alongside practical training. The practical component could leverage a model-based framework for maritime cyber risk assessment developed by the University of Plymouth, enabling officers to apply risk scenarios to real-world shipboard operations [45].

Research within the CYMET project analysed 10 different bachelor's degree programs in navigation across various European maritime universities [46]. Alarmingly, none of these programs offered specialised courses in maritime cybersecurity, despite the growing reliance on digital systems in the maritime industry. Only two programs included basic computer science courses with minimal cybersecurity content. The researchers concluded that the current offerings were insufficient, especially considering the increasing importance of cybersecurity awareness and effective cyber risk management in modern maritime operations.

A key aspect of managing cyber risks and detecting cyber incidents onboard ships is forensic readiness. As highlighted by [47], the forensic requirements of ships differ significantly from those in traditional IT systems, primarily due to the complexity of maritime operations and the unique challenges of working at sea. Unlike conventional systems, where established procedures for cyber incident detection and investigation are in place, ship crews typically receive little to no training in identifying cyber threats or handling digital evidence.

Currently, the IMO does not mandate any regulations regarding the retention of cyber-related evidence, leaving a significant gap in the maritime sector's ability to respond effectively to cyberattacks. As [31] suggest, improving forensic readiness is essential for ensuring that ships can handle the aftermath of a cyber incident, from securing evidence to investigating the root causes of attacks.

To address these gaps, [47] recommends a set of actions aimed at improving the sector's forensic readiness. These include comprehensive training programs for ship staff, crew members, and management that focus on raising awareness of cyber risks, equipping personnel with the skills to detect cyber threats, and ensuring secure handling of digital evidence. This proactive approach is crucial as maritime operations increasingly rely on digital technologies, making ships more vulnerable to cyberattacks.

Several researchers have conducted surveys among shipping companies and maritime professionals to assess the state of cybersecurity within the maritime sector [15, 48–53]. These surveys aimed to evaluate the attitudes, actions, and preparedness of companies regarding cybersecurity, while also identifying the educational and training needs of maritime personnel, both ashore and afloat.

The findings revealed that maritime systems and ships, often perceived as cyber-secure, exhibit significant vulnerabilities in critical components. Research conducted in Estonia specifically highlighted the very real and potentially damaging nature of cyber threats at sea [15]. The most reported incidents included malware infections in ship computers, phishing attacks, email spoofing, GPS interference, ransomware, and network malfunctions.

A study by [54] reinforces these findings, revealing that maritime cybersecurity preparedness is still lacking in many organisations, particularly in terms of incident response and recovery planning. The study also identified common attack vectors, including weaknesses in crew training, human error, and social engineering, further stressing the need for enhanced cybersecurity education across the sector.

Chupkemi and Mersinas [55] identified 18 key challenges related to maritime cybersecurity training and compliance. These challenges, validated through a survey of over 200 maritime professionals, point to human factors, such as limited training resources and resistance to new technologies, as the primary barriers to effective cybersecurity education. Their findings underscore the importance of integrating compliance programs with comprehensive training to address human risk in the maritime sector.

Research published by [56] adds that the increasing reliance on digital platforms and automation in ship operations has widened the attack surface for cybercriminals, making it essential for shipping companies to adopt a more structured approach to cyber risk management and align with industry best practices. Additionally, maritime-specific concerns such as GPS spoofing, interference with automated systems, and vulnerabilities in operational technology are particularly alarming for the sector.

According to the feedback from shipping companies, the most significant cyber threats originate from third parties such as hackers, suppliers, passengers, and port officials. However, internal risks were also noted, including vulnerabilities in IT systems onboard and insufficient security procedures. These findings are consistent with broader research, such as a study that highlights the growing importance of cyber risk management in maritime operations and underscores the critical role of human preparedness in mitigating cybersecurity threats [57].

The CyberOwl and HFW report highlights the significant financial impact of cyberattacks in the maritime industry, with the average ransom paid by shipowners reaching US \$3.2 million. Despite these high costs, many companies continue to under-invest in cybersecurity, with over 50% of respondents spending less than US \$100,000 annually on cyber risk management. Additionally, 25% of the survey participants indicated that they do not have insurance coverage for cyber risks, further emphasising the industry's vulnerability to cyber threats [58].

Maritime companies are increasingly recognising the importance of cyber hygiene and cybersecurity awareness training. Many have shown interest in implementing cybersecurity drills and integrating cyber incident response into their broader risk management strategies. This proactive approach is crucial, especially as cyberattacks targeting operational technologies grow more sophisticated, as highlighted by various recent reports and surveys [54, 55, 59–61].

The maritime industry, traditionally conservative, has lagged other sectors in adopting new technologies. Several studies have identified significant deficiencies in maritime education, particularly regarding seafarers' digital skills. The EU-funded project SkillSea was launched to address this issue, aiming to equip maritime professionals with essential digital, green, and soft management skills to meet the demands of the evolving maritime labour market. According to the project's reports, the maritime sector faces several key challenges in adapting to future technological advancements [62]. Research conducted within the SkillSea project revealed that over 50% of seafarers believe the STCW Convention lacks critical topics, while one-third consider the current training framework to be "overburdened with obsolete knowledge" [63]. Additionally, around 30% of seafarers indicated that STCW competencies for marine engineering and operational control are insufficient for onboard duties. About 24% identified gaps in navigation skills, and 20% pointed to inadequate training in radiocommunications. The survey also highlighted four major areas where seafarers see the most significant skill deficiencies:

- About 62% cited the need for creative thinking and problem-solving skills.
- About 61% pointed to a lack of familiarity with digital technologies, including cybersecurity.
- Around 55% noted gaps in teamwork and interpersonal relations.
- About 54% saw deficiencies in subjects related to maritime law, insurance, and P&I coverage.

As the industry increasingly relies on digital services, these skills are becoming more critical. This includes not only general digital proficiency but also the specific skills needed to maintain cybersecurity both onboard ships and in shore-based operations.

3 Cybersecurity in Maritime Education and Training

Maritime Education and Training (MET) institutions play a crucial role in shaping the future workforce of the maritime industry by equipping maritime professionals with the essential skills and knowledge necessary for their careers at sea. These institutions are integral to the educational ecosystem, providing a pathway for individuals to acquire the qualifications required for various roles within the maritime sector. Central to the MET framework is adherence to international and regional regulatory standards, notably the International Convention on Standards of Training, Certification, and Watchkeeping. This convention establishes the minimum requirements that MET programs must comply with to ensure the preparedness and competency of maritime personnel. The supervision of maritime educational institutions in the EU, in relation to STCW requirements, is primarily conducted by the member states themselves. They are responsible for ensuring that their seafarer training programs comply with the international standards set by the STCW Convention. Additionally, the European Maritime Safety Agency (EMSA) supports and oversees the member states by conducting regular inspections and assessments to ensure compliance with Directive 2008/106/EC and related international standards.

As the maritime industry continues to integrate more advanced digital technologies, the importance of cybersecurity training has become increasingly evident. However, current maritime education is lagging other sectors in addressing these new challenges. Research shows that many maritime training institutions offer little to no formal cybersecurity education, leaving seafarers underprepared for the threats they face at sea [44, 62].

The SkillSea project highlighted that more than 50% of seafarers believe critical topics such as digital and cybersecurity skills are missing from the STCW Convention. Furthermore, one-third of seafarers stated that the current training curriculum is overburdened with outdated knowledge. This suggests that the existing maritime education framework, while comprehensive in traditional safety and operational training, lacks sufficient focus on modern digital threats.

Maritime Cyber Security (MarCy) program offers a modular training approach tailored to the specific roles of seafarers and shore-based staff [64]. Developed using the Critical Events Model (CEM), MarCy provides 11 modules covering topics like basic cyber hygiene, regulatory requirements, and advanced cyber risk management. This role-based training ensures that personnel receive only the knowledge they need for their duties. Expert evaluations confirmed its effectiveness in filling gaps in maritime cybersecurity training.

The International Maritime Organization has recognised the increasing cyber threats facing the maritime sector and is working to address these challenges through education and regulatory updates. One key initiative was the adoption of Resolution MSC.428(98), introduced in 2017 [22]. While this resolution mainly targets companies, it highlights the importance of training seafarers to recognise and mitigate cyber risks in their daily operations.

To support this, cyber risk management guidelines were developed by the industry organisations (BIMCO, ICS, IUMI, OCIMF, and others) that provide high-level recommendations on maritime cyber risk management to safeguard shipping from current and emerging cyber threats and vulnerabilities and include functional elements that support effective cyber risk management. The guidelines focus on risk identification, mitigation strategies, and incident response, emphasising the critical role of human factors in managing cybersecurity onboard ships [61]. By emphasising the need for trained personnel, these guidelines underline the necessity of integrating cybersecurity training into Maritime Education and Training (MET) programs.

The IMO has also initiated steps to update the STCW Convention to include mandatory cybersecurity training. These updates will introduce specific modules on cyber risk management, digital safety, and incident response, ensuring that all seafarers receive the necessary education to tackle modern cyber threats [65]. According to the current timeline, the IMO plans to finalise these STCW updates by 2027, making cybersecurity a formal part of seafarer certification and training.

4 The Role of Behavioural Change Models in Cybersecurity Training for Future Seafarers

The maritime industry, with its increasing reliance on digital technologies for navigation, communication, and cargo management, faces growing threats from cyberattacks. The interconnected nature of maritime operations, from ship systems to port facilities, makes the industry particularly vulnerable. For future seafarers, cybersecurity is not just a technical challenge but also a behavioural one. To ensure that these future ship officers can protect their vessels and the broader maritime ecosystem, it is essential to focus on their behaviour in response to cybersecurity threats.

Cybersecurity awareness training for maritime students must move beyond theoretical knowledge and develop secure behaviour that can be sustained over time because theoretical knowledge alone does not guarantee the correct application of security measures in real-life situations. The maritime industry is unique in its operational challenges, with ships often being isolated at sea, dependent on digital systems for navigation, communication, and cargo management. A cyber incident in this context could result in severe operational and safety risks, including disruptions to navigation, engine control failures, or loss of communication with shore-based authorities.

Without developing secure behaviours, students may understand the importance of cybersecurity but fail to consistently apply best practices under pressure in maritime environment or in day-to-day operations [66]. The ability to react appropriately to the cyber threats requires habitual behaviour that has been reinforced through practical, scenario-based training. Secure behaviour must be instinctive, not just intellectual, especially in environments where quick decisions are often required.

Furthermore, the threats faced by the maritime industry are constantly evolving. Cybercriminals and hackers are continually developing new techniques, meaning the students need to not only understand the threats as they exist today but also develop the mindset and habits that will help them adapt to emerging threats in the future. Therefore, introducing long-term secure behaviours helps future ship officers remain vigilant and proactive, ensuring the safety of their vessel, crew, and the broader maritime ecosystem, even as cyber threats evolve.

This is where behavioural change models come into play. By understanding what motivates behaviour, what enables it, and what triggers action, training programmes can be designed to introduce the right cybersecurity habits in future maritime officers. Two of the most relevant models for this purpose are Susan Michie's COM-B model and BJ Fogg's behaviour model [67, 68]. Several other behavioural change models have been applied across different industries to improve cybersecurity practices.

The Health Belief Model (HBM), originally developed to explain health-related behaviours, suggests that individuals assess the benefits and barriers of taking action to avoid risks [69, 70]. However, HBM lacks a focus on the specific capabilities or environmental factors that influence behaviour, making it less comprehensive for developing consistent secure behaviour among seafarers.

The Theory of Planned Behaviour (TPB) explains that behaviour is driven by intentions, which are influenced by attitudes, social norms, and perceived control over the action [71]. For future seafarers, their intention to comply with cybersecurity protocols might be shaped by how they view the importance of these protocols, the expectations of their superiors, and their confidence in following through with secure actions.

While TPB provides insights into how students may form intentions, it does not fully address the external factors, such as training opportunities or available resources that might enable or hinder secure behaviour onboard ships.

The Transtheoretical Model (TTM), also known as the Stages of Change Model, explains that behavioural change happens gradually, moving through stages like pre-contemplation, contemplation, preparation, action, and maintenance [72]. For maritime students, this could mean recognising that their current cybersecurity practices are inadequate (contemplation), learning new skills (preparation), and then applying these skills onboard (action). While this model highlights the gradual nature of behaviour change, it does not provide guidance on how to develop the capabilities or create the environmental opportunities that make secure behaviour more likely, particularly in complex maritime environments.

The COM-B model and Fogg's behaviour model are well-suited for the maritime context because they consider not only the individual's motivations but also their abilities and the environment in which they operate. These models are practical frameworks that can help future seafarers develop and maintain secure behaviour in high-risk and high-stress maritime settings [73, 74].

The COM-B model, developed by Susan Michie, provides a holistic view of behaviour by considering capability, opportunity, and motivation. For students, this model is highly relevant as it acknowledges that secure behaviour is not only a result of knowledge but also of the environment and motivation within which they operate.

- Capability: For future seafarers, capability involves both the technical skills needed to navigate cyber threats and the cognitive skills to understand the risks. Cybersecurity training for students should include hands-on exercises where they can practice responding to cyber incidents, such as detecting malware or

responding to a phishing attempt either on the ship's network or on personal devices. Scenario-based training can build these capabilities by simulating real-world cyberattacks in a maritime setting.
- Opportunity: Opportunity involves creating an environment where secure behaviour is encouraged and possible. Onboard ships, this could mean having the right cybersecurity tools, such as regularly updated software, access to cyber threat intelligence tools, secure communication channels and file exchange, available backups, and supportive leadership that prioritises cybersecurity. Maritime training programmes must also emphasise the importance of organisational culture in fostering secure behaviour, ensuring that all crew members understand their role in maintaining cyber hygiene.
- Motivation: Motivation is crucial for encouraging seafarers to engage in secure behaviour. In the maritime industry, demonstrating the real-world consequences of cyberattacks—such as the loss of control over navigation systems or cargo handling—can be a powerful motivator. Students must see the connection between cybersecurity and the safety of the ship, its crew, and its operations. Motivation can also be enhanced through rewards, such as recognition for good cybersecurity practices or successfully completing cybersecurity drills.

BJ Fogg's behaviour model simplifies behaviour change by focusing on three elements: motivation, ability, and triggers. For maritime students, these three factors are critical in ensuring that secure behaviour is adopted and maintained.

- Motivation: Maritime students can be motivated by showing them the potential consequences of a cyber breach at sea or in port. For instance, a ransomware attack that locks down ship systems could result in not only financial losses but also physical dangers, such as collisions or cargo losses. Demonstrating these risks during training helps increase the motivation to engage in secure behaviours.
- Ability: In the Fogg model, ability refers to how easy it is to perform the behaviour. In the maritime industry, simplifying cybersecurity processes is crucial. If secure behaviour, like updating software, using two-factor authentication or using password manager, is made simple and accessible, seafarers are more likely to adopt it. Training should focus on reducing the complexity of cybersecurity tasks to ensure that students feel capable of performing them.
- Triggers: Triggers are prompts that remind students to act securely. In a ship environment, this could be as simple as automated reminders to update passwords, warnings when insecure connections are detected, or alerts that prompt the crew to check for software updates. These triggers can be integrated into the ship's daily operations, ensuring that cybersecurity becomes a routine part of life onboard.

Several other sectors have successfully integrated these behavioural change models into their cybersecurity training. In healthcare, the COM-B model has been used to help pharmacy professionals improve secure data handling and patient care practices [75]. This model focuses on capability (ensuring staff have the skills to handle data securely), opportunity (providing the necessary systems and environments for secure practices), and motivation (reinforcing secure behaviour through feedback and support). This approach has been shown to improve confidence and capability in applying learned skills, even in complex environments like community pharmacies. Maritime training can adopt similar approaches by ensuring seafarers have the technical skills and secure systems to handle sensitive ship data and communications.

In the maritime industry, where cyber threats can have significant operational and safety impacts, it is essential that future ship officers are equipped with not only technical skills but also the right behavioural habits. The COM-B model and Fogg's behaviour model provide practical frameworks for developing these habits. By focusing on capability, opportunity, and motivation, as well as simplifying secure actions and providing regular triggers, maritime cybersecurity training can prepare maritime students to navigate the cybersecurity challenges they will face at sea.

A significant challenge to cybersecurity awareness training for the maritime professionals is the persistence of a blame culture, where crew members are hesitant to report issues, errors, or potential anomalies due to fear of criticism or punishment [76]. This cultural mindset can severely impact cybersecurity, as it discourages open communication and timely reporting of potential threats. In cybersecurity, where rapid response is critical to mitigating risks, this reluctance can leave vessels vulnerable to undetected threats.

To successfully apply behavioural change models in cybersecurity awareness training for future seafarers, it is crucial to address underlying cultural barriers. Models like the COM-B and Fogg's behaviour model emphasise the need for creating the right environment and motivating individuals to adopt secure behaviours. While awareness of cyber threats is important, ENISA's findings point out that focusing solely on fear-based approaches is largely ineffective at changing cybersecurity behaviours [77]. This insight is particularly important when motivating future seafarers. Instead of focusing on the dangers alone, training should emphasise coping mechanisms and skills-building, ensuring that students feel confident in responding to cyber threats. This approach resonates with Fogg's behaviour model, which highlights that a sense of self-efficacy, the belief in one's ability to act securely, is critical in motivating behaviour change.

Therefore, training programmes should focus on building an environment where reporting potential cyber risks is seen as a positive and responsible action, rather than as a sign of failure. Moving away from a blame culture will encourage students to develop proactive reporting habits, which are critical for effectively managing cyber threats in the high-risk maritime industry.

A key element in overcoming a blame culture onboard ships is strong support from leadership. Ship officers should actively encourage crew members to report any unusual digital activity, whether it involves the ship's systems or personal devices that are being used in ship's network. By fostering open communication, leaders can create a safe environment where crew members feel comfortable reporting cybersecurity risks without fear of being blamed. The Fogg behaviour model, which focuses on motivation and making actions easy, can be applied here. When reporting is simple and reinforced with positive feedback, seafarers are more likely to follow secure practices.

Additionally, cybersecurity awareness training should include specific modules on how to report threats effectively. Incorporating these skills into regular exercises and simulated cyberattacks will help normalise reporting, making it a routine part of seafarers' duties. This approach aligns with the COM-B model, which emphasises the importance of capability, ensuring that seafarers have the skills and confidence to report cybersecurity threats.

By incorporating behavioural change models into the cybersecurity awareness training programme, it is possible to break down the traditional blame culture and promote a reporting culture. This ensures that future seafarers are not only aware of cyber threats but are also empowered to act by reporting anomalies, leading to a more secure and resilient maritime environment.

5 Case Study: TalTech Estonian Maritime Academy

Tallinn University of Technology (TalTech) is Estonia's leading institution for engineering and IT education. The Estonian Maritime Academy, a part of TalTech, is the country's only institution providing multi-level maritime education and conducting professional research in marine sciences [78]. There are four higher education programmes of applied studies, two master's and one doctoral study programme. The academy aims to train skilled maritime professionals and contribute to research and development, striving to be a centre of maritime excellence and a reliable partner in international projects.

Key features of the academy include an advanced simulator centre that offers high-tech solutions for innovative cross-border training, optimising learning time and cost. The academy employs around 100 lecturers and instructors with diverse expertise and hosts approximately 500 students across applied higher education, master's, and doctoral programs. The comprehensive training provides a 5-year study period for future navigators and ship engineers, combining academic learning with seagoing practice.

TalTech Estonian Maritime Academy recognised the growing importance of cybersecurity in the maritime industry. In the academic year 2020–2021, the academy introduced a new course titled "Introduction to Cybersecurity" as a mandatory subject for students pursuing navigation studies [79]. This initiative was inspired by research highlighting the critical need for cybersecurity knowledge among seafarers [35]. Despite cybersecurity training not being required by the STCW Convention at that time, the Estonian Maritime Academy proactively addressed this educational gap.

Initially, the six ECTS course (16 weeks) was exclusive to navigation students. However, over the subsequent years, it expanded significantly. In 2021, 18 navigation students enrolled. The following year, enrolment increased to 38. By 2023, the course became an elective for Port and Shipping Management and master's students, attracting 44 participants. In 2024, enrolment grew to 53 students from various programs. Recognising future needs, the academy made the course mandatory for ship engineering students in 2022, with the first cohort set to take it in the spring of 2025, leading to even higher enrolment compared to the previous year.

The primary goal of the course is to provide students with a solid understanding of cybersecurity principles and practical skills to manage cyber risks aboard ships. The curriculum aims to help students understand basic security concepts and terminology, recognise major cyber risks and threats to ships and maritime organisations, and become familiar with cybersecurity guidelines and legislation specific to the maritime sector. Additionally, students learn about threats to the information society, the consequences of information security breaches, and best practices in cyber hygiene. Ethical aspects of cybersecurity are also emphasised.

The course covers a wide range of topics. It begins with an introduction to the fundamentals of cybersecurity, including key terms and general concepts. Students independently complete an online cybersecurity course provided by CISCO to build foundational knowledge. The curriculum delves into cryptography and secure communication, teaching principles of cryptography, how to use PGP for secure emails, and data backup strategies. Legislation and guidelines are examined, covering national and international laws and maritime cybersecurity guidelines.

A significant portion of the course addresses cyber threats in maritime operations. Real-world cyber incidents in the maritime sector are analysed, along with vulnerabilities in IT and operational technology (OT) systems on ships. Discussions include future trends in maritime digitalisation and their implications for cybersecurity. Students receive information about cyber exercises, such as those organised by the NATO Cooperative Cyber Defence Centre of Excellence (CCDCOE), to gain an understanding of their purpose and execution.

The concept of Security Operations Centres (SOC) is introduced, highlighting their role and function in maritime and other sectors. The course also covers digital forensics and incident handling, teaching procedures for managing cyber incidents, preserving evidence, and reporting. Cyber risk management is emphasised, instructing students on integrating cyber risk into Safety Management Systems (SMS) and effectively communicating with company IT departments.

Last 2 years (2023–2024), the course has emphasised developing new habits as a core element. These habits aim to enhance both personal life and IT behaviour. Throughout the course, students are required to track their new habits daily and record their progress in a table created by the lecturer. Students choose suitable habits from a list provided by the lecturer. The habit-tracking table includes information such as the "trigger" for completing the habit that day, the student's mood and stress levels, how automatically the habit was performed, reasons why the habit wasn't completed that day, and the Self-Report Habit Index [80]. Additionally, at the beginning and end of the course, all students take the HAIS-Q test [81]. Other tests are also conducted to assess the student's knowledge of cybersecurity in the maritime sector.

Human factors and psychology are integral components of the curriculum. Topics such as awareness of social engineering tactics, overcoming a blame culture to encourage incident reporting, personal experiences, and psychological aspects of cybersecurity are explored. An innovative aspect of the course is the focus on developing new habits. Throughout the course, students choose personal and IT-related habits to adopt, such as regularly updating software or using strong passwords. They track their progress daily using habit-tracking tools, which helps to develop automatic secure behaviours that extend beyond the classroom.

To engage students, the course employs various teaching methods. Online lectures and recorded sessions using platforms like MS Teams allow all students to participate simultaneously and review content as needed. Students are required to complete the online CISCO course independently and must pass it within the first month of their study program. Group work and presentations encourage collaboration and peer learning. Reading academic articles by other researchers plays a crucial role, offering the opportunity to explore emerging trends in maritime cybersecurity. Practical exercises provide hands-on experience in tasks such as setting up secure email communication and creating data backups. Guest lectures from industry experts and cybersecurity professionals enrich the curriculum by sharing real-world insights.

Understanding that changing behaviour is key to improving cybersecurity, the course incorporates behavioural change models like the COM-B model and Fogg's behaviour model. These models help students develop long-term secure habits by focusing on building the necessary skills through practical exercises, creating an environment that supports secure behaviour, and highlighting the real-world impact of cyber threats to motivate students to adopt secure practices.

According to ENISA's Cybersecurity Culture Guidelines, one key finding is that increasing threat awareness alone is insufficient for fostering secure behaviours among workers, especially in high-risk environments like maritime operations. Instead, interventions that focus on enhancing users' capability and providing the tools to cope with cyber threats are more effective [77]. This aligns directly with the COM-B model, which stresses the importance of developing both the capability and opportunity for secure actions. For maritime students, this means practical, hands-on training, enabling them to manage cyber incidents effectively and ensuring they have access to the necessary tools and resources onboard ships to apply what they have learned.

Feedback from students has been overwhelmingly positive. They appreciated the practical focus of the course and found the real-world case studies engaging. Some initially found technical aspects challenging, like using PGP encryption, but felt that overcoming these challenges was rewarding. The habit-tracking exercise, though initially seen as time-consuming by some, ultimately helped them develop good habits. Students valued the guest lectures from industry experts, which provided insights into real-world applications of cybersecurity. Many expressed intentions to continue practicing the new habits they developed during the course.

The course has had a significant positive impact. Students have a better understanding of cyber threats and how to manage them. They have adopted new behaviours that enhance cybersecurity, both personally and professionally. The course also improved their responses to simulated phishing attacks, demonstrating enhanced incident recognition and appropriate reactions. The increasing enrolment indicates the growing relevance and importance of cybersecurity in maritime professions.

Conducting such a course requires solid knowledge of both cybersecurity and maritime operations. Since the field is very broad, several guest lecturers from Estonia and abroad, who are experts in their field, have been invited to provide students with their years of experience in a way that is engaging and offers practical value. Additionally, collaboration with local institutions provides necessary information, such as insights on cyber incidents and expectations that future employers have for their new hires.

Furthermore, online classes sometimes serve as a meeting point for students, guest lecturers, and representatives from maritime companies. In such cases, even representatives from ports or shipping companies benefit by gaining new insights into cybersecurity developments from researchers or learning about the challenges faced by penetration testers who assess the IT and OT systems of ships, ports, or offshore platforms. These interactions foster engaging discussions that provide valuable learning opportunities for students.

Since the course is important for all young people entering the maritime sector, interest in it has only increased. In addition to the teaching materials currently used in the classes (such as cybersecurity guidelines, Hak5 hacking tools, etc.), there are plans to make greater use of the Wärtsilä ship bridge simulator at the maritime academy and to implement the maritime cyber lab, which is currently under development as part of the EU funded Horizon 2020 project MariCybERA [82].

5.1 Enhancing Cybersecurity Awareness Through the Level of Paranoia (LoP)

Level of Paranoia (LoP) refers to a heightened state of vigilance that maritime students are encouraged to cultivate during cybersecurity training. One of the key goals of this training is to raise the LoP, enhancing students' awareness and caution when interacting with digital systems. The intention is to help them avoid careless errors without introducing excessive fear or mistrust. In cybersecurity, especially in high-risk environments like maritime operations, maintaining an elevated LoP is essential for preventing threats such as phishing, ransomware, and system breaches. Through scenario-based exercises, students are trained to approach digital interactions with healthy scepticism, encouraging them to pause, question, and verify actions that might expose vessel systems to cyber threats. Similar methods have been applied in other sectors, such as healthcare, where heightened alertness is critical for managing high-stakes situations. This increase in LoP helps students develop alertness, allowing them to make quick but careful decisions in the fast-paced and high-pressure maritime environment.

6 Challenges in Developing Cybersecurity Awareness Course for Maritime

Developing and implementing such a cybersecurity awareness course for maritime students have its own challenges. The biggest issue may be the lack of specialists. Teaching cybersecurity for maritime sector requires experts who have both a technical background in cybersecurity and knowledge of the maritime industry. Finding such specialists can be difficult, especially considering that many experts prefer to work in the private sector or as consultants, where salaries are higher than in the education system. Additionally, instructors need to continuously train and update their skills, as the field of cybersecurity evolves rapidly. Offering training opportunities and staying up to date with new threats can be resource- and time-consuming.

Curricula and training materials for maritime cybersecurity are still in the development phase. Most traditional cybersecurity programs are focused on land-based systems and may not meet the specific needs of the maritime sector. Maritime training institutions need to invest in their own tailored educational programs or develop them in collaboration with other organisations specialising in IT security. The learning materials being developed must cover various aspects of cybersecurity, such as the protection of ship ICT systems, information management, cybersecurity procedures and controls, incident management, and cyber risk assessment.

In addition, a challenge in teaching the new course lies in the fact that traditional seafarer training has been very hands-on and focused on mechanical and operational skills. Most maritime students may not have a strong background in IT, making it difficult to grasp cybersecurity concepts. Furthermore, some future seafarers may view cybersecurity as something that doesn't fall directly within their job responsibilities, leading to low motivation to learn it. Raising awareness and changing the mindset about the importance of cybersecurity in their daily work are crucial but a challenging task.

The lack of international standards for cybersecurity training for seafarers makes it difficult to create unified and widely recognised educational programs. Maritime academies often must develop their own programs without clear guidelines, which means that the quality and content of the courses can vary.

If the course is mandatory for both navigation and ship engineering students, it is important to consider that due to their different roles, they face different cybersecurity threats. Navigators are responsible for the ship's navigation and communication systems, while engineers work with operational systems and automated technology. For example, navigators need knowledge about disruptions in navigation systems and the risks of cyberattacks that affect the ship's manoeuvrability, whereas engineers need practical skills in protecting the automation and control systems used to maintain the ship's operations. Cybersecurity training must be role-specific, but developing such a differentiated program can be complex and time-consuming.

In conclusion, TalTech Estonian Maritime Academy's approach to integrating cybersecurity into maritime education demonstrates a proactive response to the challenges of the digital age. By focusing on practical skills, behavioural change, and collaboration with industry experts, the academy prepares future seafarers to navigate the evolving cyber threats in maritime operations. This case study highlights the importance of adapting maritime education to include cybersecurity, ensuring the safety and security of maritime activities in the digital era.

7 Conclusion

The maritime industry's shift towards digitalisation has greatly improved connectivity and operational efficiency. However, this technological progress has also opened the door to numerous cyber threats, posing serious risks to the safety and security of maritime operations. This chapter highlights the urgent need to rethink seafarer training to address these new cybersecurity challenges.

The analysis shows a significant gap between the rapid digital transformation of maritime activities and the current level of cybersecurity awareness among seafarers. The absence of comprehensive cybersecurity provisions in the STCW Convention, along with inconsistent application of IMO guidelines, highlights the necessity for fundamental changes in maritime education and training.

By introducing behaviour change models like the COM-B model and Fogg's behaviour model, we can bridge this gap effectively. These models focus on developing secure behaviours by enhancing individuals' capabilities, providing opportunities, and boosting motivations to adopt and maintain best practices in cybersecurity. Integrating these models into maritime curricula allows training programmes to move beyond mere theoretical knowledge, fostering lasting behavioural changes among future seafarers.

The case study of TalTech Estonian Maritime Academy demonstrates the effectiveness of this approach. By making cybersecurity training compulsory and embedding behaviour change principles into their programme, the academy has successfully cultivated a security-conscious culture among its students. Positive feedback from students, increased cybersecurity awareness, and the adoption of secure habits all attest to the programme's success.

In conclusion, strengthening the maritime industry's resilience against cyber threats depends on proactively adopting innovative training methods that prioritise behavioural change. Maritime education institutions need to involve this new approach to equip future seafarers with not only the necessary knowledge but also the established practices required to navigate the complexities of the digital age. This commitment is essential for safeguarding maritime operations, protecting critical infrastructure, and ensuring the safety of vessels, crews, and cargo in today's increasingly interconnected world.

Acknowledgments Research for this publication was funded by the EU Horizon2020 project 952360-MariCybERA.

References

1. United Nations Conference on Trade and Development, *Review of Maritime Transport 2021* (United Nations Publications, 2021), https://unctad.org/webflyer/review-maritime-transport-2021. Accessed 22 Aug 2022
2. United Nations Conference on Trade and Development, *Review of Maritime Transport 2023* (United Nations Publications, Geneva, 2023), https://unctad.org/system/files/official-document/rmt2023_en.pdf
3. N. Gardner, M. Kenney, N. Chubb, *A Changed World* (Thetius, 2021)
4. F. Alqurashi et al., *Maritime Communications: A Survey on Enabling Technologies, Opportunities, and Challenges*, https://www.researchgate.net/publication/360233150
5. A.Y. Cil, D. Abdurahman, I. Cil, Internet of Things enabled real time cold chain monitoring in a container port. J. Ship. Trade **7**(1), 9 (2022)
6. D. Gavalas, T. Syriopoulos, E. Roumpis, Digital adoption and efficiency in the maritime industry. J. Ship. Trade **7**, 11. https://doi.org/10.1186/s41072-022-00111-y
7. V. Karetnikov, E. Ol'Khovik, A. Ivanova, A. Butsanets, Technology level and development trends of autonomous shipping means, in *Energy Management of Municipal Transportation Facilities and Transport*, (Springer, 2019), pp. 421–432
8. A. Greenberg, *The Untold Story of NotPetya, the Most Devastating Cyberattack in History* (Wired, 2018)
9. C. Shen, Cosco Shipping targeted in ransomware attack. Lloyd's List. https://lloydslist.maritimeintelligence.informa.com/LL1123581/Cosco-Shipping-targeted-in-ransomware-attack. Accessed 12 Sept 2022
10. A. Tsonchev, Troubled waters: cyber-attacks on San Diego and Barcelona's ports—Darktrace Blog, https://darktrace.com/blog/troubled-waters-cyber-attacks-on-san-diego-and-barcelonas-ports. Accessed 13 Sept 2022
11. UN Maritime Agency hit by 'sophisticated cyberattack'—SecurityWeek, https://www.securityweek.com/un-maritime-agency-hit-sophisticated-cyberattack/. Accessed 7 Sept 2024
12. Naval Dome exposes vessel vulnerabilities to cyber attack, https://www.seatrade-maritime.com/maritime-safety/naval-dome-exposes-vessel-vulnerabilities-to-cyber-attack. Accessed 7 Sept 2024
13. D. Heering, O.M. Maennel, A.N. Venables, Shortcomings in cybersecurity education for seafarers, in *5th International Conference on Maritime Technology and Engineering 2020*, (2020)
14. İ. Karaca, Ö. Söner, An evaluation of students' cybersecurity awareness in the maritime industry. Int. J. 3D Print. Technol. Digit. Ind. **7**(1), 78–89 (2023). https://doi.org/10.46519/ij3dptdi.1236264
15. D. Heering, Ensuring cybersecurity in shipping: reference to Estonian shipowners. TransNav **14**(2), 271 (2020)
16. Cyber security survey shows more action is needed in the industry, https://www.bimco.org/news/priority-news/20180924-cyber-security-survey. Accessed 7 Sept 2024
17. Maritime Cyber Priority 2023: staying secure in an era of connectivity, https://www.dnv.com/cybersecurity/cyber-insights/maritime-cyber-priority-2023/. Accessed 8 Sept 2024
18. A. Olsen, *Cyber Security Characteristics of the Maritime Industry* (Springer, 2024), pp. 577–592. https://doi.org/10.1007/978-3-031-55943-3_41
19. Seafarers win commitment to mandatory internet access in international law | ITF Seafarers, https://www.itfseafarers.org/en/news/seafarers-win-commitment-mandatory-internet-access-international-law. Accessed 8 Sept 2024
20. *New Important Set of Amendments to the MLC, 2006 Will Enter into Force on 23 December 2024* (International Labour Organization), https://www.ilo.org/resource/news/new-important-set-amendments-mlc-2006-will-enter-force-23-december-2024. Accessed 9 Sept 2024

21. Sign the petition for free internet at sea—SEAFiT, https://seafit.safety4sea.com/petition/. Accessed 8 Sept 2024
22. International Maritime Organization, *Resolution MSC.428(98) Maritime Cyber Risk Management in Safety Management Systems* (International Maritime Organization, 2017), https://wwwcdn.imo.org/localresources/en/OurWork/Security/Documents/Resolution%20MSC.428(98).pdf. Accessed 22 Sept 2024
23. D. Cimpean, J. Meire, V. Bouckaert, V.C. Stijn, A. Pelle, L. Hellebooge, *Analysis of Cyber Security Aspects in the Maritime Sector* (ENISA, 2011)
24. Ø.J. Rødseth, H.-C. Burmeister, Risk assessment for an unmanned merchant ship. TransNav **9**(3), 357–364 (2015). https://doi.org/10.12716/1001.09.03.08
25. O. Fitton, D. Prince, B. Germond, M. Lacy, *The Future of Maritime Cyber Security* (Lancaster University, 2015)
26. J. DiRenzo, D.A. Goward, F.S. Roberts, The little-known challenge of maritime cyber security, in *2015 6th International Conference on Information, Intelligence, Systems and Applications (IISA)*, (IEEE, 2015), pp. 1–5. https://doi.org/10.1109/IISA.2015.7388071
27. J. Bhatti, T.E. Humphreys, Hostile control of ships via false GPS signals: demonstration and detection. Navigation **64**(1), 51–66 (2017). https://doi.org/10.1002/navi.183
28. D. Bothur, G. Zheng, C. Valli, A critical analysis of security vulnerabilities and countermeasures in a smart ship system, in *Proceedings of the 15th Australian Information Security Management Conference*, vol. 2017, (AISM, 2017), pp. 81–87
29. M. Fruth, F. Teuteberg, Digitization in maritime logistics—what is there and what is missing? Cogent Bus. Manage. (2017). https://doi.org/10.1080/23311975.2017.1411066
30. A. Garcia-Perez, M. Thurlbeck, E. How, *Towards Cyber Security Readiness in the Maritime Industry: A Knowledge-Based Approach* (Semantic Scholar, 2017), pp. 1–7
31. K.D. Jones, K. Tam, M. Papadaki, Threats and impacts in maritime cyber security. Eng. Technol. Ref. **2016**, 1–12 (2016). https://doi.org/10.1049/etr.2015.0123.Published
32. K. Kolev, N. Dimitrov, Cyber threats in maritime industry-situational awareness and educational aspect, in *Global Perspectives in MET: Towards Sustainable, Green and Integrated Maritime Transport*, (2017), pp. 352–360
33. Y.-C. Lee, S.-K. Park, W.-K. Lee, J. Kang, Improving cyber security awareness in maritime transport: a way forward. J. Kor. Soc. Mar. Eng. **41**(8), 738–745 (2017). https://doi.org/10.5916/jkosme.2017.41.8.738
34. I. Radmilo, A. Gudelj, P. Ristov, Information security in maritime domain, in *International Maritime Science Conference*, (Faculty of Maritime Studies Split, Solin, 2017), pp. 76–93
35. D. Heering, Ensuring cyber security in shipping with reference to estonian shipowners and proposals for risk mitigation, Dissertation, Tallinn University of Technology, Tallinn, 2017, https://digikogu.taltech.ee/et/item/7bb85829-2c56-4c8e-9895-f955385f627b. Accessed 9 Sept 2024
36. S. Ahvenjärvi, Addressing cyber security in training of the mariner of the future—the CYMET project, in *International Symposium on Integrated Ship's Information Systems & Marine Traffic Engineering Conference*, (2018)
37. M.A. Alfultis, Educating the future maritime workforce in a sea of constant disrupters and change, in *AGA 2018—19th Annual General Assembly (AGA) of the International Association of Maritime Universities (IAMU)*, (2018), pp. 87–93
38. O.S. Hareide, O. Josok, M.S. Lund, R. Ostnes, K. Helkala, Enhancing navigator competence by demonstrating maritime cyber security. J. Navig. **71**(5), 1025–1039 (2018). https://doi.org/10.1017/S0373463318000164
39. R. Kidd, E. Mccarthy, Maritime education in the age of autonomy. WIT Trans. Built Environ. **187**, 221–230 (2019). https://doi.org/10.2495/mt190201
40. J. Mileski, C. Clott, C.B. Galvao, Cyberattacks on ships: a wicked problem approach. Marit. Bus. Rev. **3**(4), 414–430 (2018). https://doi.org/10.1108/mabr-08-2018-0026
41. K. Tam, K.D. Jones, Maritime cybersecurity policy: the scope and impact of evolving technology on international shipping. J. Cyber Policy **3**(2), 147–164 (2018). https://doi.org/10.1080/23738871.2018.1513053

42. R. Zăgan, G. Raicu, R. Hanzu-Pazara, S. Enache, Realities in maritime domain regarding cyber security concept. Adv. Eng. Forum **27**, 221–228 (2018). https://doi.org/10.4028/www.scientific.net/aef.27.221
43. M.S. Lund, O.S. Hareide, Ø. Jøsok, An attack on an integrated navigation system. Sjøkrigsskolen **3**(2), 149–163 (2018). https://doi.org/10.21339/2464-353x.3.2.149
44. L. Roolaid, Küberturbe haridus laevaohvitseride väljaõppes ning soovitused selle korraldamiseks, Dissertation, Tallinn, 2018, https://www.ester.ee/record=b5382048*est. Accessed 9 Sept 2024
45. K. Tam, K. Jones, MaCRA: a model-based framework for maritime cyber-risk assessment. WMU J. Marit. Aff. **18**, 129 (2019). https://doi.org/10.1007/s13437-019-00162-2
46. S. Ahvenjärvi et al., Safe information exchange on board of the ship. TransNav **13**(1), 165–171 (2019). https://doi.org/10.12716/1001.13.01.17
47. K. Tam, K.D. Jones, Forensic readiness within the maritime sector, in *2019 International Conference on Cyber Situational Awareness, Data Analytics and Assessment (Cyber SA)*, (2019), pp. 1–4. https://doi.org/10.1109/CyberSA.2019.8899642
48. J.I. Alcaide, R.G. Llave, Critical infrastructures cybersecurity and the maritime sector. Transport. Res. Proc. **45**, 547–554 (2020). https://doi.org/10.1016/j.trpro.2020.03.058
49. P. Bolat, G. Yüksel, S. Uygur, A study for understanding cyber security awareness among Turkish seafarers, in *Second Global Conference on Innovation in Marine Technology and the Future of Maritime Transportation*, vol. 8, (2016), p. 479. https://doi.org/10.1007/s11628-013-0202-1
50. P. Bolat, G. Kayişoğlu, Antecedents and consequences of cybersecurity awareness: a case study for Turkish maritime sector. J. ETA Marit. Sci. **7**, 344 (2019). https://doi.org/10.5505/jems.2019.85057
51. A. Dimakopoulou, N. Nikitakos, I. Dagkinis, T.E. Lilas, D.A. Papachristos, M. Papoutsidakis, The new cyber security framework in shipping industry. J. Multidiscip. Eng. Sci. Technol. **6**(12), 11227–11233 (2019)
52. K. Tam, K.D. Jones, Situational awareness: examining factors that affect cyber-risks in the maritime sector. Int. J. Cyber Situat. Awaren. **4**(1), 40–68 (2019). https://doi.org/10.22619/IJCSA.2019.100125
53. K. Tam, K.D. Jones, Factors affecting cyber risk in maritime, in *2019 International Conference on Cyber Situational Awareness, Data Analytics and Assessment (Cyber SA)*, vol. 2019, pp. 1–8. https://doi.org/10.1109/CyberSA.2019.8899382
54. How to improve the cyber security awareness in the shipping industry, Degree Programme in Maritime Management (2023)
55. D.C. Chupkemi, K. Mersinas, Challenges in maritime cybersecurity training and compliance. J. Mar. Sci. (2024). https://doi.org/10.3390/jmse12101844
56. M. Haugli-Sandvik, M.S. Lund, F.B. Bjørneseth, Maritime decision-makers and cyber security: deck officers' perception of cyber risks towards IT and OT systems. Int. J. Inf. Secur. **23**(3), 1721–1739 (2024). https://doi.org/10.1007/S10207-023-00810-Y
57. I. Moen, A. Oruc, A. Amro, V. Gkioulos, G. Kavallieratos, Survey-based analysis of cybersecurity awareness of Turkish seafarers. Int. J. Inf. Secur. **23**, 3153–3178 (2024). https://doi.org/10.1007/s10207-024-00884-2
58. CyberOwl, HFW report: maritime industry pays average US$3m ransom in cyberattacks, HFW, https://www.hfw.com/about-us/news/cyberowl-hfw-report-maritime-industry-pays-average-us3m-ransom-in-cyberattacks-march-2022/. Accessed 22 Sept 2024
59. The Great disconnect: the state of maritime cyber risk management, Thetius, https://thetius.com/the-great-disconnect-the-state-of-maritime-cyber-risk-management/. Accessed 22 Sept 2024
60. Maritime forecast to 2050 by DNV, https://www.dnv.com/maritime/publications/maritime-forecast/. Accessed 22 Sept 2024
61. The guidelines on cyber security onboard ships, https://www.bimco.org/about-us-and-our-members/publications/the-guidelines-on-cyber-security-onboard-ships. Accessed 22 Sept 2024

62. D. Zec, L. Maglic, H.M. Šimić, A. Gundić, *Current Skills Needs: Reality and Mapping* (SkillSea, 2020)
63. ETF: European Transport Workers' Federation, *Radical Changes Required to Future-Proof Training and Education of Maritime Professionals* (ETF: European Transport Workers' Federation), https://www.etf-europe.org/radical-changes-required-to-future-proof-training-and-education-of-maritime-professionals/. Accessed 22 Sept 2024
64. A. Oruc, N. Chowdhury, V. Gkioulos, A modular cyber security training programme for the maritime domain. Int. J. Inf. Secur. **23**(2), 1477–1512 (2024). https://doi.org/10.1007/s10207-023-00799-4
65. Working at sea: IMO review of STCW identifies new focus areas, IMCA, https://www.imca-int.com/working-at-sea-imo-review-of-stcw-identifies-new-areas-of-focus/. Accessed 22 Sept 2024
66. K. Tam, R. Hopcraft, T. Crichton, K. Jones, The potential mental health effects of remote control in an autonomous maritime world. J. Int. Marit. Saf. Environ. Affairs Ship. **5**(2), 40–55 (2021). https://doi.org/10.1080/25725084.2021.1922148
67. S. Michie, M.M. Van Stralen, R. West, The behaviour change wheel: a new method for characterising and designing behaviour change interventions. Implement. Sci. (2011). https://doi.org/10.1186/1748-5908-6-42
68. B.J. Fogg, *A Behavior Model for Persuasive Design* (2009), www.bjfogg.com
69. H. Anuar, S.A. Shah, H. Gafor, I. Mahmood, H.F. Ghazi, Usage of Health Belief Model (HBM) in health behavior: a systematic review. Malay. J. Med. Health Sci. **16**(Suppl 11), 201–209 (2020)
70. I.M. Rosenstock, Historical origins of the health belief model. Health Educ. Behav. **2**, 328 (1974)
71. I. Ajzen, From intentions to actions: a theory of planned behavior, in *Action Control: From Cognition to Behavior*, ed. by J.K. Julius, Beckmann, (Springer Berlin Heidelberg, Berlin, 1985), pp. 11–39. https://doi.org/10.1007/978-3-642-69746-3_2
72. J.O. Prochaska, W.F. Velicer, The transtheoretical model of health behavior change. Am. J. Health Promot. **12**(1), 38–48 (1997). https://doi.org/10.4278/0890-1171-12.1.38
73. J. Rengamani, M.S. Murugan, A study on the factors influencing the seafarers' stress. AMET Int. J. Manag. **4**(1), 44–51 (2012)
74. M. Oldenburg, H.-J. Jensen, Maritime field studies: methods for exploring seafarers' physical activity. Int. Marit. Health **70**(2), 95–99 (2019)
75. E.M. Seston et al., Implementation of behaviour change training in practice amongst pharmacy professionals in primary care settings: analysis using the COM-B model. Res. Soc. Adm. Pharm. **19**(8), 1184–1192 (2023). https://doi.org/10.1016/J.SAPHARM.2023.04.123
76. M. Carrera Arce, R. Baumler, Effective learning from safety events reporting takes two: getting to the root & just culture. TransNav **15**(2), 331–336 (2021). https://doi.org/10.12716/1001.15.02.08
77. ENISA, *Cybersecurity Culture Guidelines: Behavioural Aspects of Cybersecurity* (ENISA, 2018). https://doi.org/10.2824/324042
78. EMERA Study Programmes, TalTech, https://taltech.ee/en/emera-study-programmes. Accessed 22 Sept 2024
79. TalTech ÕIS, http://ois2.taltech.ee/uusois/subject/VLL1480. Accessed 28 Sept 2024
80. B. Verplanken, S. Orbell, Reflections on past behavior: a self-report index of habit strength. J. Appl. Soc. Psychol. **33**(6), 1313–1330 (2003). https://doi.org/10.1111/J.1559-1816.2003.TB01951.X
81. K. Parsons, A. Mccormac, M. Butavicius, M. Pattinson, C. Jerram, Determining employee awareness using the Human Aspects of Information Security Questionnaire (HAIS-Q). Comput. Secur. (2014). https://doi.org/10.1016/j.cose.2013.12.003
82. Centre for Maritime Cybersecurity, TalTech, https://taltech.ee/en/estonian-maritime-academy/areas-of-advance/maritime-cyber-security. Accessed 28 Sept 2024

Open Access This chapter is licensed under the terms of the Creative Commons Attribution-NonCommercial-NoDerivatives 4.0 International License (http://creativecommons.org/licenses/by-nc-nd/4.0/), which permits any noncommercial use, sharing, distribution and reproduction in any medium or format, as long as you give appropriate credit to the original author(s) and the source, provide a link to the Creative Commons license and indicate if you modified the licensed material. You do not have permission under this license to share adapted material derived from this chapter or parts of it.

The images or other third party material in this chapter are included in the chapter's Creative Commons license, unless indicated otherwise in a credit line to the material. If material is not included in the chapter's Creative Commons license and your intended use is not permitted by statutory regulation or exceeds the permitted use, you will need to obtain permission directly from the copyright holder.

Ports of Tallinn and Koper Comparative Analysis Including Cybersecurity

Sanja Bauk, Bojan Beskovnik, and Seçil Gülmez

1 Introduction

As digitalization increases, so does the threat of cyber-attacks. This applies to all areas of life and business. Shipping and port logistics are no exception. The maritime economy and industry involve complex activities, so the possibilities for cyber-intrusions are many and varied. Researchers in this field are looking for appropriate protection measures and mechanisms. These are at the intersection of human factors, cybersecurity awareness, and skills of ports' employees, seafarers, and other involved parties. Here, it is important to consider technology vulnerabilities related to port infrastructure, including manned and unmanned vessels that the port services, as well as related internal and external cyber-security attack vectors and countermeasures. Seaports are "microcosm" of maritime transportation system (MTS). Any vulnerability in any segment of the MTS can be found at a port. Concerning cyber safety and security, it is to be highlighted that emerging cyber regulations focus on operational technology (OT), but most attacks affect information technology (IT). Although OT and IT are converging, most port security managers come from a physical security, and most ports do not have a Chief Information Security Officer (CISO) [1].

Some port authorities have reacted to cyberthreats in a proactive manner. For instance, the first U.S. port to open a Cybersecurity Operations Center (CSOC) was the Port of Los Angeles in 2014. The CSOC is certified to comply with the International Organization for Standardization (ISO) 27001 information security

---

S. Bauk (✉) · S. Gülmez
Estonian Maritime Academy, Tallinn University of Technology, Tallinn, Estonia
e-mail: sanja.bauk@taltech.ee; secil.gulmez@taltech.ee

B. Beskovnik
Faculty of Maritime Studies and Transport, University of Ljubljana, Ljubljana, Slovenia
e-mail: bojan.beskovnik@fpp.uni-lj.si

© The Author(s) 2025
S. Bauk (ed.), *Maritime Cybersecurity*, Signals and Communication Technology,
https://doi.org/10.1007/978-3-031-87290-7_4

standard [2]. The port proposed a Cyber Resilience Center in early 2019, signed an IBM contract in late 2020, and opened for operation in early 2022. To support its stakeholders, which include ocean carriers, terminal operators, freight and cargo haulers, and other members of the maritime supply chain, this second-generation maritime cyber facility is designed to offer threat detection, attack analysis, information sharing, and threat response strategies.

Furthermore, the Maritime Cybersecurity Operations Center (MSOC) was established by the Maritime and Port Authority (MPA) in Singapore in May 2019. The center's mission is to protect maritime critical information infrastructure from potential cyber-attacks by providing early detection, monitoring, analysis, and response. There is also a regional Information Fusion Center (IFC) run by the Republic of Singapore Navy. Established in 2009, the IFC unites delegates from over 20 countries to oversee maritime security incidents, encompassing everything from illicit fishing and human trafficking to piracy and smuggling. In 2020, the IFC saw the addition of cybersecurity as a topic of interest.

Here is a summary of the most significant cyber-attacks that have been reported on ports worldwide between 2012 and 2021 to highlight the necessity of protecting the port from cyber-attacks (Table 1).

2 Environmental Scan

Digitalization is increasingly present in shipping. It is also present in ports, which act as hubs between sea, land, and air transport. As a result, all aspects of port operations will be at the forefront of the industry's digital transformation. As digitalization increases, so does the risk of cyber-attacks. One of the most important tasks is to build a bridge between theory and practice in this area. The knowledge gained should be used to bring the maturity of the system to a proactive and predictive level and to continuously improve security in the ports. Transport companies and port stakeholders in the maritime industry are at different stages of digital transformation. While the most successful examples of digital transformation can be seen in highly digitalized ports and companies (such as the ports of Singapore and Rotterdam), many other ports and port stakeholders and companies involved in maritime supply chains are lagging behind [7]. While many attacks have been reported in recent years and cybersecurity is a concern for port operators and practitioners, the scientific community has not paid much attention to the topic and has produced few scholarly contributions. Following an evaluation of current regulations, legislators should introduce and enact warning systems and procedures to respond to cyber incidents. They should also create a new organizational framework for assessing these risks, encourage more research, and, from a transnational perspective, facilitate cooperation between all parties involved in the global supply chain [8]. Port technologies allow port officers to perform a variety of tasks and manage them from a remote-control station. These operations include, but they are not limited to cargo handling, control of automated vehicles and

Table 1 Some seaports' cyber-attacks [3, 4]

Year	Port	Attack	Description
2012	Australian Customs and Border Protection Service across terminals in Brisbane, Sydney, Melbourne, and Fremantle	Hackers broke into a cargo management system	The criminal organization was transporting contraband goods in cargo containers mixed with legitimate cargo
2011–2013	Port of Antwerp, Belgium	Broke into the port's computers	Cybercriminals were working with drug and arms smugglers. The smugglers recruited hackers for social engineering, physical access to network devices, and using of snooping devices (e.g., keystroke loggers). The aim was sending bogus bills of lading to the port for the containers with contraband goods
2012–2014	Danish Port Authority	Virus spread through the Authority network	The Danish Maritime Authority was attacked in 2012 by a PDF (Portable Document Format from Adobe) containing a virus. The virus spread throughout the Danish Maritime Administration's network and Danish government organizations before it was discovered in 2014 [5]
2017	Several ports worldwide for which Maersk shipping company is responsible	NetPetya ransomware attack; distributed denial of service (DDoS)	The staffers began seeing messages advising them that their file systems were being repaired and messages that some important files had been encrypted. A demand was made for $300 million in Bitcoin to obtain the encryption key
2018	Port of Longview, on the Columbia River in Washington	Intrusion detected by FBI	This attack affected personal information for 370 past and current employees, as well as 47 vendors
2018	Port of Barcelona	Ransomware attack	Several servers in the port were breached, but there was no impact on port operations
2018	Port of San Diego	SamSam ransomware attack	The attack affected more than 500 workers, and disrupted computer systems, business services, and Harbor Police operations

(continued)

Table 1 (continued)

Year	Port	Attack	Description
2020	Port of Kennewick, on the Columbia River in Washington	Ransomware that bypassed firewalls and antimalware software and lock system administrators out of their systems	The perpetrators demanded a $200,000 ransom, while all port services were down for several days as servers were re-built from offline backups
2020	Iranian Port facility attack; Shahid Rajaee terminal, near the Iranian coastal city of Bandar Abbas on the Strait of Hormuz	Computer strike	Computer strike was carried out presumably by Israeli operatives, in retaliation for an earlier attempt to penetrate computers that operate rural water distribution systems in Israel [6]
2021	Port of Houston	Web server suffered a root-level intrusion by an unidentified state actor	A Zero Day attack, where the attackers were able to install malicious code
2021	Ports of Cape Town and Durban	Death Kitty ransomware attack	More than 1 TB of Transnet's corporate files, including financial records were encrypted; container operations were disrupted and web sites inaccessible for several days

different handling equipment, entrance procedures, optical character recognition on vehicles, scanning systems (X-ray and gamma ray), and hybrid electronic technologies, including electronic tracking and tracing services via various wireless devices and networks. Port operators must have specific knowledge and skills to fulfil these tasks successfully. Senarak [9] specified 24 knowledge and 26 skills indicators, which he divided into four groups: technology and network cyber risk prevention, cybersecurity and threat management, cyberthreat and security management, and information and system management. In particular, the connected technologies have hurriedly ushered ports into the digital era, necessitating the acquisition of digital skills and knowledge by port personnel now more than ever. The results indicate that there is currently a significant gap in the literature because real-time data on maritime cyber-attacks are lacking, which makes it challenging to model and predict such attacks in the future. Furthermore, there has not been much discussion of the financial effects of maritime cyber-attacks. The capacity of the curriculum and educational system to prepare maritime professionals to counter present and emerging cyber threats is limited. Besides, the existing legal frameworks at the national and international levels are insufficient to effectively govern the maritime cyberspace. Senarak [10] investigated the relationships between port cybersecurity organization and cyber threats and he proposed three levels of the port cybersecurity organization model based on human, infrastructure, and procedure factors to help policymakers in developing cybersecurity measures. Cyber terrorism may arise from human weakness, while cyber criminality may result from inefficient infrastructure. If the factor of the port's cybersecurity procedures was poorly implemented, container ports were likely to be affected by cyber espionage. Therefore, to ensure a culture of cyber threat awareness at all organizational levels, it is essential that all port employees, including executives and manager levels are trained and educated. Munyai and Govender [11] recommend to enhance human resources and cybersecurity tools to limit unauthorized access to sensitive business information and maintain its security. Moreover, stringent adherence to ISPS Code protocols and other preventive measures is also necessary. On top of this, the legal framework is crucial for safeguarding against cyber-attacks. The insufficiency of International Maritime Organization (IMO) legal instruments pertaining to maritime cybersecurity was brought to light by Karim [12]. A framework for the cybersecurity of ships and ports has been established by the IMO Guidelines and Resolution MSC.428(98) in the margins of the ISM and ISPS Codes. While it is overshadowed by a legal instrument such as the SOLAS Convention, the existing IMO initiatives offer much good and are an important factor in raising awareness of cybersecurity among stakeholders. It also calls on national maritime administrations to provide measures that cyber risks are adequately managed in port security systems. The resolution makes it mandatory under the ISM Code to incorporate cybersecurity into the current safety management system. Nevertheless, the port ecosystem still has a lot of space for strategic, tactical, and operational advancements in both preventing and responding to cyber-attacks. Given that ports and terminals are vital infrastructures that are necessary for the local, regional, and international economies, policymakers should take a serious look at this problem and collaborate with the industry to guarantee the highest level of cyber protection.

3 Problem Statement

This study deals with the topic of cybersecurity in seaports. After the introductory section, we have provided an overview of recent major cyber-attacks on ports around the world. These and similar attacks should be a lesson in how important the fight against cyber-attacks is. We have given a brief overview of the relevant literature sources in this field and then we have focused on the experiences related to cybersecurity in two ports in the European Union (EU): the ports of Tallinn (Estonia) and Koper (Slovenia). We have described the ports in terms of their traffic volume and capacity, focusing on container, Ro-Ro and passenger flows, as these flows can be categorized as more vulnerable as they are more complex to monitor and identify the potential threat. This is especially true for containers, as each shipment has a different shipper, consignee, transport route, etc. We also described the level of digitalization in the ports and their roles, i.e., the services they offer to terminal operators, freight forwarders, passengers, and other stakeholders. We found that both ports are very vulnerable to cyber-attacks due to their high level of digitalization. Our goal was to identify cyber threats and vulnerabilities. Using this information, we aimed to create a risk assessment matrix based on our own knowledge and the experience of the experts we interviewed. To this end, we analyzed relevant academic and media literature and conducted in-depth interviews with key information security officers in the ports of Tallinn and Koper.

4 Methodological Approach

Considering potential cyber risks, we examined the volume of traffic and the degree of digitalization at the ports of Tallinn and Koper. We used these ports' annual reports for this purpose. Additionally, we conducted a desktop literature review for this study using ScienceDirect and ProQuest literature sources in relation to the environmental scan. Then, we identified cyber threats and vulnerabilities matrix, based on [13]. Afterwards, we conceived the interview and survey questions for the ports' IT/OT security officers, who are experts in the field of cybersecurity in the considered ports. These questions were sent to two external experienced colleagues in research methodology to evaluate them. Their valuable suggestions in refining the questions were taken into the account. Based on our knowledge, experiences, and logical framework we employed, we have categorized each question into the groups of potential cybersecurity risks (Table 2) to better link the questions with the threat and vulnerability matrix later. The applied approach was mixed one, based on the qualitative analysis of the interview questions and quantitative analysis

Table 2 Matrix of cyber-threats and vulnerabilities in the port

Threats	Vulnerabilities
T1. Terrorist attack T2. Sabotage T3. Malicious destruction of data and facilities	Lack of physical security Lack of a uniform physical security policy enforcement Lack of environmental protection Inadequate policy and procedures for physical and environmental security Inadequate monitoring of the facilities Lack of logical access security Inadequate audit logs to detect unauthorized access of the premises Lack of a formal procedure to verify the access rights of employees on the premises No uninterruptible power supply system No power conditioning equipment The site is located in an area prone to power outages The site is located in an environmentally susceptible area (electronic interference, extreme temperatures, humidity, pollution, etc.) Insufficient monitoring of environmental conditions No specific allocation of roles and responsibilities in relation to continuity or disasters Inadequate procedures for managing changes to infrastructure components, etc.
T4. Denial of service T5. Unauthorized software changes T6. Unauthorized data access T7. Web site intrusion	Inappropriate assignment of roles and responsibilities in relation to continuity or disasters Missing formal/informal disaster and recovery plans Missing business continuity plans for the data recovery Inadequate recovery procedures Improper or inadequate technical facility maintenance Inadequate backup policy Inadequate backup processes for software and data Unavailable backup files and systems Lack of backup facilities and processes Lack of maintenance of equipment and facilities Lack of careful planning and routing of cables Lack of cryptographic means for the data integrity protection, etc.
T8. Theft and fraud T9. Failure of outsources operations T10. Vermin (adware, malware, phishing, pop-ups, spyware, viruses, Trojans, and worms)	Inadequate network management (routing resilience) Missing documents for policies and procedures for physical control of software and hardware Lack of a firewall Improper or inadequate maintenance of technical equipment Missing documents and test security plans to protect systems and networks Lack of a formal process to review employee access rights to the premises Lack of a comprehensive security awareness and training programme, etc.

Adapted from [13]

Table 3 Risk assessment matrix for a seaport environment (own research)

		Severity				
		Not significant	Minor	Moderate	Major	Severe
Likelihood	Almost certain					
	Likely		T10	T8		
	Possible			T4, T5, T6	T3, T7	
	Unlikely					
	Rare			T9	T2	T1

of some of the survey questions. In the Appendix at the end of the chapter are given both interview and survey questions. Having in mind that we have had in-depth interviews and administered surveys with only two specialists in charge of information safety and cybersecurity for the ports under consideration, the collected information was manually cleaned up, edited, coded, and analyzed. The following sections encompass the results and conclusion remarks.

We did our best to identify some common threats and vulnerabilities regarding information safety and cybersecurity in the considered ports of Tallinn and Koper with reference to [13] propositions. Namely, based on the threats identified, we have proposed a risk assessment matrix (Table 3; see Table 2 as well). This matrix is based on the extensive literature review and on our experience but may be subject to further investigation and adjustments.

According to our assessment, there are no "red" threats, but there is also only one "green" threat. Most of the issues analyzed are in the "orange" and "yellow" zones. For example, we have assessed the failure of outsourced operations as rare but with medium severity for port operations if the port outsources some IT/OT services. The yellow zones contain most of the identified threats, including terrorist attacks, sabotage, denial of service, unauthorized software modifications, unauthorized data access, website intrusions, malicious destruction of data and facilities, and malware (adware, malware, phishing, pop-ups, spyware, viruses, Trojans, and worms). The orange-colored zones are, in our estimation, occupied by threats such as theft and fraud, denial of service, and malicious destruction of data and facilities. We have proposed this matrix to give professionals an idea of how to access the threats and to give researchers a basis for upcoming more comprehensive investigations.

The relationships between the analyzed factors of cybersecurity in the port container terminal and the questions asked in the interviews and surveys are shown in Table 4. The criteria relevant to cybersecurity in the port were determined after a detailed review of the relevant literature. On the other hand, the questions were formulated after taking note of the recommendations in [1, 11]. The interview and survey questions can be found in the appendices (Interview Questions and Survey Questions) at the end of the manuscript.

Below we have described the ports of Tallinn and Koper in terms of turnover and level of deploying IT/OT solutions. We then conducted interviews via MS Teams with two information safety and cybersecurity experts in these ports and sent out

Table 4 Criteria versus interview and survey content (own research)

Main criteria	Sub-criteria	Description	Interview	Survey
Cybersecurity Governance and Policy Framework	Existence and enforcement of comprehensive cybersecurity policies	Ensures that an organization has documented cybersecurity policies that guide actions and decisions, providing a clear roadmap for managing cyber risks	Q1	Q2, Q3, Q4, Q6, Q8, Q9, Q11
	Regular review and updates of cybersecurity policies	Reflects the dynamic nature of cyber threats, ensuring that policies remain relevant and effective against evolving risks		
	Defined governance structure with clear roles and responsibilities	Establishes a framework for accountability and decision-making in cybersecurity efforts, clarifying who is responsible for what actions		
	Compliance with international standards and guidelines	Demonstrates adherence to widely recognized cybersecurity best practices and benchmarks, enhancing the organization's cybersecurity posture		
Risk Assessment and Management	Implementation of risk assessment methodologies	Involves systematic processes to identify, evaluate, and prioritize risks based on their potential impact, ensuring that resources are allocated effectively to mitigate threats	Q1, Q2, Q4, Q5, Q6	Q2, Q3, Q4, Q5, Q7, Q8, Q9, Q11
	Regular updates of risk management strategies	Ensures that risk management approaches adapt to new threats, technologies, and changes in the operational environment		
	Critical asset identification and protection	Involves identifying assets that are vital to the organization's operations and applying protective measures to safeguard them from cyber threats		
	Incident impact analysis and management	Focuses on understanding the potential consequences of cybersecurity incidents and developing strategies to minimize their impact on operations		

(continued)

Table 4 (continued)

Main criteria	Sub-criteria	Description	Interview	Survey
Technology and Infrastructure Security	Network security and management	Ensures the integrity, confidentiality, and availability of data across networks, protecting against unauthorized access and cyber-attacks	Q1, Q3	Q2, Q3, Q4, Q7, Q8, Q10
	System and application security	Involves securing software and hardware against vulnerabilities and threats, including regular updates and patches		
	IoT and sensor security	Addresses the security challenges posed by interconnected devices and sensors, ensuring their resilience against cyber-attacks		
	Data encryption and protection	Utilizes encryption and other protective measures to secure data, both at rest and in transit, against unauthorized access and breaches		
Cybersecurity Awareness and Skill Development	Training programs for maritime personnel	Provides maritime staff with the knowledge and skills to recognize and respond to cybersecurity threats, promoting a secure operational environment	Q1, Q2, Q7	Q1, Q2, Q3, Q4, Q5, Q6, Q9
	Specialized training for cybersecurity roles	Develops the expertise of individuals in key cybersecurity positions, ensuring that they are equipped to manage and mitigate cyber risks effectively		
	Cyber hygiene and best practices dissemination	Encourages the adoption of good cybersecurity practices among all personnel, reducing the risk of cyber incidents due to human error		
International Collaboration and Regulation	Engagement in international cybersecurity frameworks	Participating in global cybersecurity initiatives to share knowledge, align with best practices, and strengthen collective cyber resilience	Q1, Q7	Q2, Q3, Q4, Q6, Q8
	Collaboration on maritime cybersecurity initiatives	Working with international partners to address shared cyber threats and challenges, enhancing security across the maritime sector		
	Harmonization of cybersecurity standards	Aims to standardize cybersecurity practices across borders, facilitating a more coherent and effective approach to cyber risk management		

Incident Response and Recovery	Formal incident response plans	Establishes predefined procedures for responding to cyber incidents, minimizing impact and facilitating a swift return to normal operations	Q1, Q2, Q4, Q5, Q6	Q2, Q3, Q4, Q5, Q8, Q9, Q11
	Regular testing and updating of response plans	Ensures that incident response plans are effective and current, preparing organizations to handle cyber incidents efficiently		
	Training and drills for staff on cybersecurity incidents	Prepares staff to respond appropriately to cyber incidents, reducing potential damage and improving recovery times		
Innovation and Technological Adaptation	Investment in emerging cybersecurity technologies	Focuses on adopting new and advanced technologies to strengthen cybersecurity defenses and stay ahead of potential threats	Q1, Q3, Q7	Q2, Q3, Q4, Q8, Q10
	Adaptation to new threats and technological advancements	Ensures that cybersecurity strategies and tools evolve in response to the changing landscape of cyber threats and the introduction of new technologies		
	Exploration of autonomous operation security measures	Investigates security considerations and measures related to the increasing use of autonomous systems in maritime operations, addressing unique vulnerabilities and challenges		
Supply Chain and Third-Party Risk Management	Cybersecurity assessments for third-party vendors	Evaluates the cybersecurity practices and vulnerabilities of third-party vendors and partners, ensuring they meet the organization's security standards	Q1, Q7, Q8	Q2, Q3, Q4, Q6, Q7, Q8, Q9, Q11
	Security controls and continuous monitoring of third-party risks	Implements controls and monitoring practices to manage and mitigate risks associated with third-party engagements		
	Secure supply chain practices	Ensures the security of the supply chain, protecting against cyber threats that can propagate through interconnected networks and partners		

questionnaires by email. The results of the study are presented in the following sections.

5 The Port of Tallinn

The Port of Tallinn is developing a balanced business model with a strong customer focus. It carries out several activities in parallel, including passenger and cargo services, shipping services, and supports real estate development in the immediate hinterland of the port. It is also a part of the offshore wind farms project in terms of providing a maintenance vessel service. The port is a constituent of a green shipping corridor—a climate neutral ferry route for passengers travelling between Estonia and Finland. In 2023, it has 8 million passengers, 564 ferry calls, 13 million tons of cargo, and 1380 cargo ships calls. It provides services to passengers and cargo traffic, and development of port infrastructure for both cargo handling and its transportation. It has capacities to provide services to dry bulk, containers, Ro-Ro, liquid bulk, and general cargo. The port aims to make its services smarter and more efficient while considering the impact on the environment. This includes economic and social impacts. More specifically, the port looks at innovation, R&D-based development, sustainable business, attractiveness as an employer, better quality of public space and regional development, health and safety, corporate social responsibility, energy efficiency, sustainable consumption, a clean Baltic Sea, increased recycling, clean air, water, and soil [14].

5.1 Port Throughput and Digitalization

Due to the sanctions against Russia and the general economic downturn, cargo handling in the Port of Tallinn approximately decreased from 18 million tons in 2022 to 13 million tons in 2023. This is the lowest level of the throughput in recent decades. Liquid bulk decreased in 2023 by 3.5 million tons, dry bulk by 0.8 million tons, and Ro-R by 0.5 million tons compared to the past year. The volume of pulpwood and steel products fell, resulting in a decrease of 0.2 million tons in breakbulk cargo. Despite a 13% decline in volume in 2023 after 9 years of growth, Ro-Ro cargo remains the most common type of cargo. Most of the Ro-Ro growth is reflected in the revenues of the passenger segment. Ro-Ro cargo is mainly transported on the routes between Estonia and Finland (Tallinn-Helsinki and Muuga-Vuosaari), and most of it is carried by passenger ferries on the Helsinki route (Fig. 1).

The decline in sales, along with the unfavorable conditions on the financial markets led to a 38% drop in profit. The profit of the Port of Tallinn fell from 25.6 million euros in 2022 to 15.9 million euros in 2023 (Fig. 2).

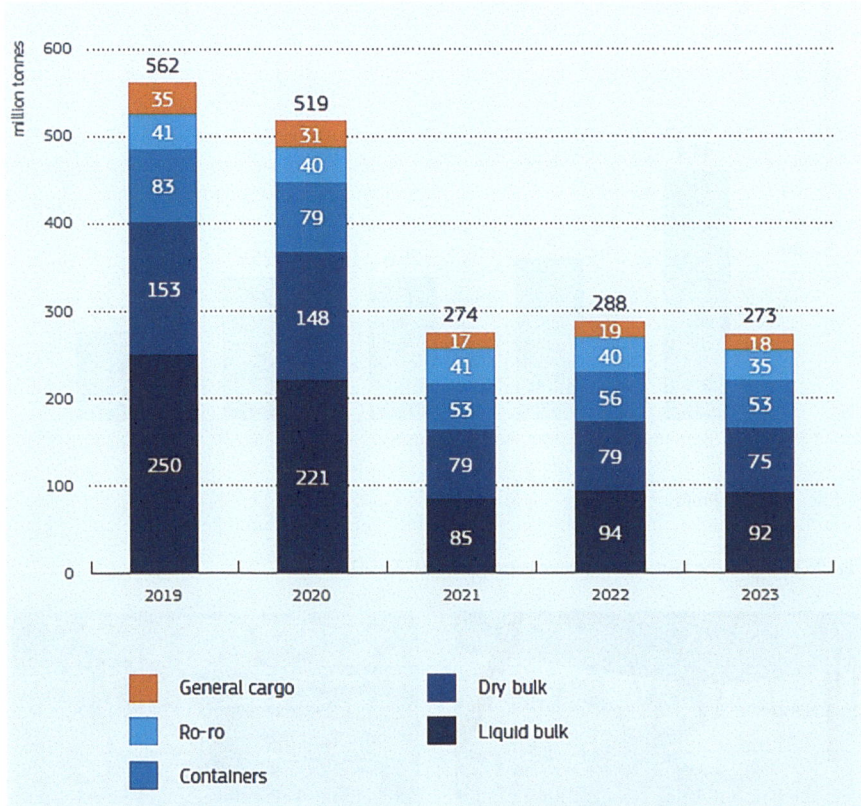

Fig. 1 Cargo volume of the largest ports on the eastern coast of the Baltic Sea, excluding Russia from 2021 [3]

The general economic downturn and the Russia-Ukraine conflict, which had a major influence on the transit of cargo to and from Russia, had the greatest effects on cargo transport last year [14]. Despite the unfavorable market situation and the crisis caused by the wars in Europe and Asia, the Port of Tallinn is working hard to improve the quality of services and the user experience, with a focus on innovation, the green agenda, and digitalization.

Passengers/Ro-Ro Terminal The Port of Tallinn is working toward a single data entry point to improve the customer experience for passengers and shippers. Electronic boarding signs, simple check-in procedures, shorter waiting times, and efficient vehicle and truck queues are all provided by the port [15]. The Port of Tallinn's Old City harbor is the fourth busiest passenger port in North Europe [16]. This port is conceived and developed as a smart one. Smart port is an intelligent system for passengers, which automates the vehicles traffic and shortens the time passenger cars and trucks spent on harbor premises. It enables automatic detection,

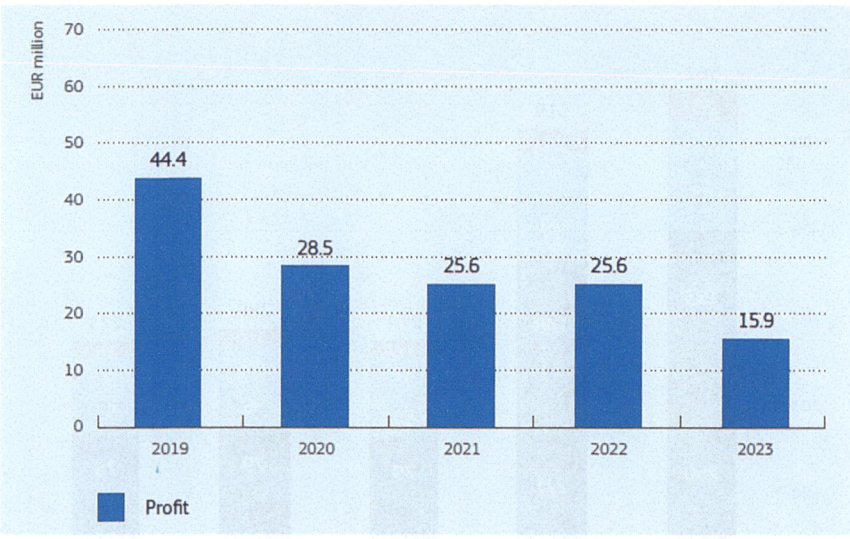

Fig. 2 Profit of the Port of Tallin per year 2019–2023 [14]

Fig. 3 Taking snapshots of the vehicle and reading its key physical parameters (**a**); digital displays for guiding the vehicles to the right gathering zone and waiting line (**b**) [16]

guidance, and check-in and loading processes of vehicles. Smart port is a globally unique concept, which automates vehicles traffic (Fig. 3).

This concept of a smart passengers/Ro-Ro terminal combines various technologies, starting with number plate recognition, the barrier and automation system, various sensors, and a check-in system that guides passengers and drivers to the assembly and (dis)embarkation areas. Such a port has a significant impact on the environment, as vehicles spend less time on the port site with their engines turned on. This reduces emissions. Smart system enables the recognition of vehicle license plates, both front and rear, pre-gate check-in, and guidance to the assembly area. The system measures the weight, height, width, and length of the vehicle and takes snapshots from different angles. This is important for safety and loading. It enables the correct positioning of the vehicles on the ferries. This information is then sent to the port management server. If the vehicle has a booking and all the information is correct, it is routed to the automatic check-in line. If not, the

Table 5 Port of Tallinn container terminals' integral features [16]

Feature	Value
Area	55 ha
Sheltered warehouses	8000 m^2
Reefer containers	404 units
Vessel loading/unloading	55–60 units/h
Queues total length	1 km
Trains traffic	4 per day

vehicle is routed to the manual check-in line. The vehicles are routed via intelligent LED displays. By automating these processes, the port saves around 320,000 labor hours per year. All of this optimizes the port's operating costs and simplifies the passenger experience by avoiding queues, reducing congestion and improving the overall logistical process.

Container Terminals Muuga Harbor and Paldiski South Harbor are the two harbors in the Port of Tallinn where containers are handled. With a capacity of 600,000 TEU annually [17], the Muuga Harbor free zone is home to the most advanced container terminal in the Baltic States region. Muuga Harbor is in an ideal location for a container terminal, with easy access to highways, trains, and logistical hubs in the hinterland. A larger free zone allows for more accommodating customs processes. Additionally, the free zone contains an industrial park that permits the construction of distribution centers. A mobile crane in Paldiski South Harbor can lift to 100 tons, which is used for loading and unloading containers. A considerable number of containers are loaded onto roller trailers and cassettes and transported by Ro-Ro ships through Paldiski South Harbor. The terminals key features are shown in Table 5.

The timetable of cargo ships in the ports of Muuga and Paldiski can be viewed on the web portal of the Port of Tallinn. However, actual arrival times may change depending on the requirements of the shipowners, terminal operators, or the port. The largest container, Ro-Ro, and general cargo terminal in Estonia is operated by HHLA TK Estonia, which was established in 1996. HHLA TK Estonia's state-of-the-art facility is situated in the Muuga Harbor free trade zone, which is easily accessible by water and land. This attracts the good's flows across international borders. To promote seamless and fast trade, HHLA TK Estonia offers its customers a comprehensive range of services for the management of local and transit cargo via the port in accordance with the widely accepted single-window system [18]. The terminal's management aims to guide development in a manner like Estonia's post-independence management, moving directly from analogue to fiber, while bypassing some development stages [19]. Besides, Estonia and the German autonomous truck developer Fernride are testing a special kind of autonomous port truck at Muuga Harbor, and if the tests are successful, the Muuga project has the potential to completely change how port operations are organized [20]. Fernride's autonomous port trucks are fully integrated into the routine port operations at the terminal, which serves as a real-world test site. The remotely controlled trucks transport containers from the terminal to the ship and back again alongside conventional, human-driven trucks.

6 Port of Koper

The Port of Koper is managed and operated by Luka Koper Ltd. Co., which was granted a 35-year concession to manage the port in 2008. In 2021, the regulation on the management of the cargo Port of Koper, performance of port activities, granting of a concession, which came into force in July 2008, was amended to increase the concession area of the port by more than 36 hectares (21 hectares for the maritime part and 15 hectares for the land part of the port) and by 11.6 hectares in 2022. The total port area measures 288 hectares, of which 117 hectares are open storage areas and 51.8 hectares are closed warehouses. The quayside is 3475 m long, with berths in three basins. There are 38 km of railway tracks in the port area, connecting each terminal with the Koper freight station.

The Port of Koper is the most important port and logistics center in Slovenia and is situated at the very northern point of the Adriatic. The strategic location of the port is important for the national economy and at the same time provides employment opportunities and contributes to the growth of regional and national trade. From an international perspective, it represents an important maritime and intermodal gateway for the markets in Central Europe.

The Port of Koper handles a variety of cargo types, including containers, automobiles, liquid and dry bulk, and general cargo. It offers a wide range of cargo handling, warehousing, and logistics services and places great emphasis on customer satisfaction by providing customized logistics solutions that meet the specific requirements of its customers and are delivered in a timely and reliable manner. Luka Koper Ltd. Co. operates 12 specialized terminals, each of which is designed for the efficient handling of different types of cargo. These terminals are divided into five profit centers according to their operational and technological characteristics. Among them, the Container Terminal (CT) is particularly noteworthy for its technical, technological, and developmental advances. It is one of the largest and most important container terminals in the Adriatic region, although it is relatively small on a global scale. The Ro-Ro terminal is also very important at European level, as it is the largest in the Mediterranean, ahead of Barcelona and Valencia, which were years ahead.

6.1 Port Throughput and Financial Data

In 2023, the Port of Koper handled 22.3 million tons of cargo, a decrease of 5.8% compared to 2022 (Fig. 4). The decline is due to general cargo handling, which fell to 1.1 million tons. There were also less dry bulk and liquid cargo, as 5.29

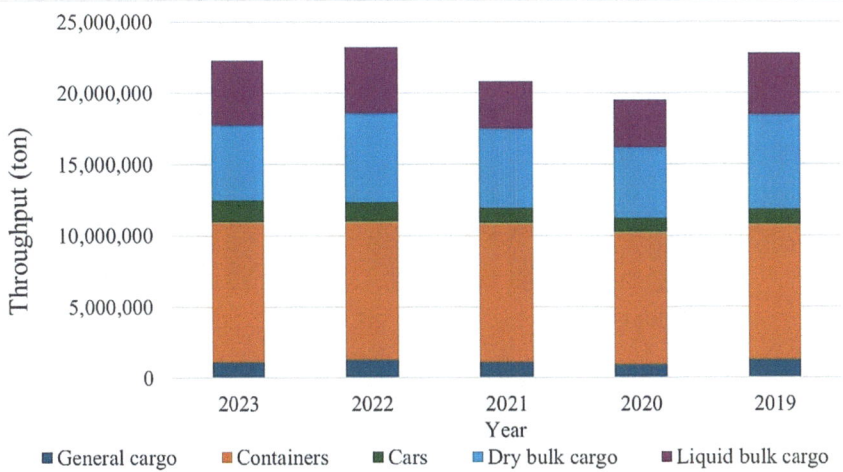

Fig. 4 Port throughput in 2019 till 2023 by type of commodity [21]

million tons of dry bulk cargo and 4.5 million tons of liquid cargo were handled. However, growth was achieved in two key segments of port development, namely containers and cars. Container handling totaled 1.066 million TEU, an increase of 4.7% compared to 1.017 million TEU in the previous year. Growth in vehicle handling was exceptionally high. With 916,728 vehicles handled, this segment grew by 14.5%. Namely, in 2022, 801,036 vehicles were handled. A total of 1642 ships arrived at the port, a similar volume to 2022. Additionally, 20,609 trains arrived at and departed from the port, with 262,583 wagons being unloaded and loaded. The share of goods transported by rail was 52%, and the share of goods transported by road was the remaining 48%, with 422,811 trucks entering the port. Net revenue in 2023 remained almost at the same level as in 2022 at EUR 312.8 million. The EBITDA fell by 18% to EUR 93.7 million in 2023. The EBITDA margin fell to 30%. The EBIT was also lower in 2023 at EUR 60.9 million. Investments totaled EUR 41.5 million, a decrease of 18% compared to the previous year [21].

Container, Ro-Ro, and Passenger Terminal Container, Ro-Ro, and passenger terminals can be categorized as more vulnerable. They are traversed by complex cargo and passenger flows. This is particularly pronounced in container transport, as each container is a separate shipment and special security procedures are required to recognize the potential threat. The container terminal covers 27 hectares and has 694.5 m of operational shoreline. The storage area measures 18 hectares, which can store around 20,700 TEU in the offshore part of the terminal and 15,000 TEU in the empty container depot. The maximum sea depth at berth is 14.5 m, which allows the berthing of ships with a capacity of 15,500 TEU. There are a total of 11 STS cranes, 4 super post-panamax, 4 post-panamax, and 3 panamax at the berths. For the transshipment in the hold, there are 27 RTG cranes, while 4 RMG cranes are used for the transshipment on the handling rails. The terminal has 5 handling tracks

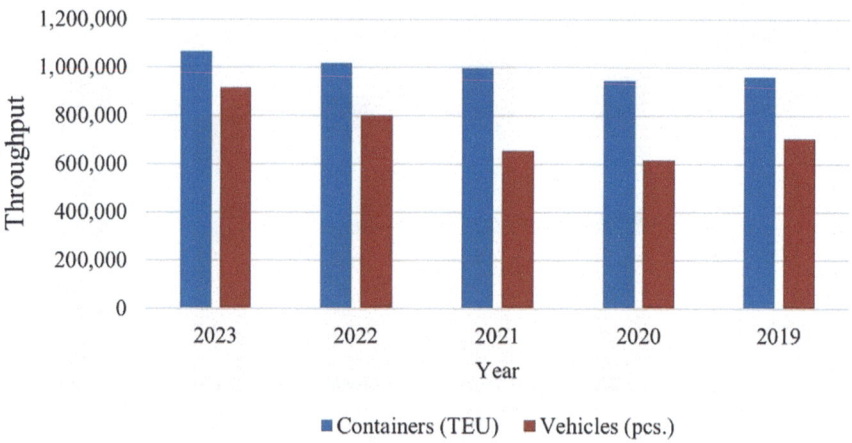

Fig. 5 Container and Ro-Ro terminal throughput in 2019 till 2023 [22]

of 700 m each. Trucks with containers and vehicles can enter the port through three gates (gate 1 close to the city center, gate 2 in Sermin, and gate 3 in Bertoki).

Container throughput is increasing. In 2023, 1,066,096 TEU were handled (Fig. 5). The 52% of the total TEUs were discharged, i.e., 552,629 TEUs. Compared to 2019, throughput in 2023 increased by just over 11%. In addition to basic handling from ship-terminal-vehicle and vice versa, several different services are offered, such as stripping and stuffing of containers, cleaning, repairing, and pre-trip inspecting of containers. Specialized services are also offered for reefer containers and in combination with general cargo terminal stuffing and un-stuffing of perishable goods. For reefer containers, they offer 24-h container monitoring, the repair of containers, and the management of power generators (maintenance, assembly, and disassembly).

A wide range of services is also offered at the Ro-Ro terminal. This terminal has 840 m of operational waterfront, with six berths and five Ro-Ro ramps. The terminal offers 67 hectares of open storage space, where 34,000 vehicles can be parked. In addition, there are 22.5 hectares of covered storage areas where 10,000 vehicles can be parked. As rail transport is developing rapidly, the terminal has 15 tracks with a total length of 7.1 km. In addition to loading and unloading services, the terminal also offers its customers vehicle washing, pre-delivery inspections, repair services, and services for the installation of additional equipment.

In 2023, 916,728 vehicles were handled at the Ro-Ro terminal (Fig. 5). A larger proportion of the vehicles are loaded onto an RO-RO ship in the port. In 2023, this amounted to 555,195 vehicles, i.e., 60% of the terminal's throughput. Compared to 2019, when 705,993 vehicles were handled, the terminal achieved growth of 30%. In 2019, the ratio between inbound and outbound vehicles was almost equal, with 48% of vehicles being inbound. The terminal is thus increasingly focusing on the export direction of European vehicles production, which come from factories in Hungary, Slovakia, the Czech Republic, and Germany.

In addition to the container terminal and the Ro-Ro terminal, it is worth mentioning the passenger terminal, which is developing rapidly and is subject to higher security measures, as over 120 thousand passengers enter Slovenia via the port of Koper every year. The terminal has a 420 m long quay and is located in the very south of the port, close to the city center. The terminal does not currently have an operational terminal building, but it is due to be built by 2028. As part of this, new security measures and equipment will be installed as the cruise market is expected to continue to grow [23].

6.2 Security Measures

The security measures in the Port of Koper are regulated by the Rules on Internal Order at Luka Koper, which have been in force since April 2011. The port has established a security system for the entire port area, which is managed by Luka Koper Ltd. Co. The following physical security services are constantly provided by port security specialists and by the national authorities: access control for all vehicles and vessels upon arrival at the port, including intrusion prevention system, 7/24 video surveillance of the port area boundaries, and fire and accident prevention. Only persons who have applied for and received a security pass in advance are authorized to enter the port. The passes are issued to employees of the port, external users, providers of agreed services and visitors and are non-transferable. Permits are also issued for motor vehicles, which are linked to the owner of the vehicle. The personal check at the entrance to the port includes a check of the validity of the pass, and a search of personal belongings, a search of clothing, and a search of the vehicle interior may also be ordered [24].

There are special restrictions in the border crossing areas by sea, which include a 10-m-wide space next to the berth of an international ship and 50 m seaward. Authorized persons with legitimate reasons are allowed to stay there. Only ships that have received prior authorization from the Slovenian Maritime Administration may enter the port. Ships may only moor at the designated berth unless there are labor requirements or safety/security circumstances.

Specific measures are aimed at anti-terrorist measures, which are in accordance with the SOLAS Convention, the ISPS Code, and the Regulation of the European Parliament and the Council of Europe on increased protection of vessels and ports. Photography of infrastructure and goods is prohibited in the port. Exceptions must be requested from the security service. It is forbidden to bring weapons, explosives, and dangerous substances into the port. Such entry also requires a permit, which is usually issued to authorized persons from the state authorities.

Cybersecurity is also important for the smooth operation of the port and the security of vehicles, people, and cargo. This refers to the transfer of information between stakeholders via the Port Community System (PCS) and other telecommunication links such as banking, e-mail, etc. To develop and test the operation of cybersecurity, several international projects have been carried out in recent years, such as 5G-LOGINNOV, Infrastress, and LUKA.DT. The 5G-LOGINNOV project has temporarily set up and tested a 5G network in the port. In collaboration with actual IT, a replacement data center was set up and tested at a different location as part of a project approach. This is a backup protocol in the event of a failure of the primary data center in Koper to ensure the smooth operation of the port. As part of the LUKA.DT project, a 3D model of the port was created to improve video surveillance of the port and ensure a higher level of security processes. The Infrastress project worked with partners to test cyber threat scenarios for liquid cargo passing through the port, that is stored in the port's hinterland. This expands the scope of cybersecurity in the port, also in line with the requirements of Directive (EU) 2022/2555 of the European Parliament and of the Council on measures for a high common level of cybersecurity across the EU.

7 Findings from the Interviews and Surveys

After a detailed analysis of the available data in the businesses and security measures related to IT/OT operations and cybersecurity in the ports of Tallinn and Koper, the results of interviews and surveys with two experts leading the information safety and cybersecurity teams in these ports are analyzed and presented in the following sub-sections of the work.

7.1 The Interview Analysis

A Cybersecurity Expert at the Port of Tallinn The interviewed expert told us that his job description includes everything related to the implementation of cybersecurity standards, or rather information security standards, the handling of information security incidents and information security education. When asked about the biggest challenges in his job, the respondent said that it was the lack of human resources. Administrators are usually busy, and it takes some time for some of them to accept information security management and actions as a part of their regular job. In response to our question about the level of digitalization in the port and the correlation between digitalization and the risk of cyber-attacks, the expert told us that the level of digitalization in the port is very high. He exposed that the Port of Tallin is a smart port. They have digital displays with automatic signs for cars and passengers; in terms of which lane to follow (i.e., where to go), automatic check-in and therefore smooth embarkation and disembarkation. They have automatic docking for some ships, while the contractors handle the cargo in the port. When asked what model of cybersecurity management they use, he replied: "Mostly in-house, while we outsource some services, such as penetration testing of some systems in the isolated environment." To our question: how do you check the cybersecurity risk level of the port?—Through internal or external audits? he replied: "Through both internal and external ones. We do everything in accordance with the Estonian Information Security Standard (E-ITS)." The E-ITS is a basis for dealing with information security. The standard is written in Estonian and is compatible with the Estonian legal system. It corresponds to the internationally recognized ISO/IEC 27001 standard for information security management [25]. When we asked about initial reports and recovery measures, the expert explained that they inform the Estonian Information Systems Authority in real time if something happens. They use a common platform for immediate reporting. If an incident could have legal implications, they report it to the police. They have regular reporting twice a year when they prepare and submit comprehensive reports. They avoid reporting small incidents as these can statistically reduce the real indicators of the level of security. When we asked him how he is informed about cybersecurity risks and mitigation measures; through private companies, national bodies, professional bodies, chambers of commerce, industry associations of professional bodies (such as IMO, EMSA, etc.), the Internet, the press and/or scientific publications, he replied: "All together." He has mailing lists, shipping companies provide him and his colleagues the information on technological developments, etc. Besides, there is a very good cybersecurity community in Estonia through which they share relevant information and knowledge.

A Cybersecurity Expert at the Port of Koper According to the Port of Koper's cybersecurity expert, the most important cybersecurity topics in the port are education, compliance with regulations, the Internet of Things (IoT) segment, and the implementation of additional measures for users and outsourcing partners. The biggest challenges in securing the port's cybersecurity processes were identified as orchestrating all solutions, optimizing incident response times, implementing new security measures for outdated solutions, and improving cybersecurity threats regarding information sharing with other ports, terminals, and critical infrastructure companies. The expert considered the port to be highly digitalized, but with a lot of room for improvement. In his opinion, security is an important part of the digitization process due to the high availability of the services offered and sometimes even the lack of alternative non-digitized procedures. The port uses a combination of in-house and outsourced cybersecurity management. The port conducts regular penetration tests, social engineering exercises, and external and internal audits to assess the level of cybersecurity risk. Cybersecurity incidents are reported to the port management and the national cybersecurity body "SICERT" (only if an incident has an impact on critical infrastructure services). Risk assessment is carried out on an annual basis. Reporting to the executive board is weekly. The cybersecurity expert is informed about cybersecurity risks and mitigation measures through the national unit SICERT—IOC MISP integration, Security Operational Center providers, security solution provider mailing lists, threat intelligence sources, and Internet OSINT tools. According to the expert, the port has physical security of data centers, access control management (MFA, PAM, etc.), staff awareness (training and internal rules), backup management, regular security audits, encryption capabilities, IT risk management, antivirus, antimalware, antispam solutions, etc.

7.2 The Survey Analysis

A Cybersecurity Expert at the Port of Tallinn The Port of Tallinn has a high level of digitization; on a scale of 1–5, the expert rated it 4. On the same scale, his rating of port employees' knowledge of cybersecurity was 3. When asked how much attention the port pays to cybersecurity, he gave the maximum estimate. He expects the number of cyber-attacks to increase in the future as digitalization increases. The main reasons are political (e.g., war in Ukraine), financial, the maliciousness of hackers, or a combination of all of these. The main problems with cybersecurity are the lack of knowledge and willingness of stakeholders to adopt new technologies, as well as the lack of legislation and regulation, due to his opinion. When asked how he would rate the risk of a cyber-attack on disruption of ship and/or fleet operations, theft of ship/cargo, damage to port/cargo handling equipment, harm to the environment, a ground of a vessel, ship collision, and physical injury or loss of life, on a scale of 1–5, he gave a score of 1 in all cases. He mentioned that level of ships autonomy is still at relatively low level. Consequently, it can be concluded that the level of risk of a cyber-attack on the Port of Talin is under constant control and therefore the danger is reduced. The port has a well-defined cybersecurity plan with more than 1500 measures, some of which are ready for implementation and some of which are in the preparatory phase. The following security schemes are applied in the port: ISO 9001, ISO 27001, NIS2 directive (extended version; based on German standards), Estonian Information Security Standards (EISS), and Information Technology Infrastructure Library (ITIL). The interviewer is familiar with the following standards ISA/IEC-62443, ISO/IEC 17799:2005, ISO 20858, ISO/IEC 27001, Information Security Forum, CobIT and Information Technology Infrastructure Library (ITIL). All of this is evidence of the port's good preparedness in terms of cybersecurity prevention and response.

A Cybersecurity Expert at the Port of Koper The Port of Koper has a well-defined cybersecurity management plan. The following security schemes are covered under the Cybersecurity Plan: ISPS, ISO 9001, SOLAS/MARPLE standards (more on the physical side), including customs and financial authority regulations. The respondent is familiar with security standards such as ISO/IEC 17799:2005, ISO/IEC 27001, CobIT, and Information Technology Infrastructure Library (ITIL). According to the respondent, there will be a rise in maritime cyber-attacks in the future. Concerning the query: how likely is each of the (listed in the survey) outcomes in the event of a cyber-attack on the Port of Koper? He came up with the following assessment on the scale 1–5: theft of ship/cargo (4), damage to port/cargo handling equipment (4), harm to the environment (3), physical injury or loss of life (2), and disruption of ship and/or fleet operations (1). On a scale of 1–5, he rated the level of overall awareness of cyber risks in the Port of Koper as 4. On the same scale, he rated the level of cybersecurity knowledge of the port employees at 4. He rates the Port of Koper's attention to cybersecurity as very high (4 out of 5).

8 Conclusions

The Ports of Tallinn and Kopar are in different geographical locations, function in different geo-political environments, and have similar yearly throughput but they achieve different performances in container and passenger business. Both ports are highly digitalized and therefore exposed to a significant degree to the risks of cyber-intrusions. In discussions with the main information safety and cybersecurity experts at these ports, we found that both ports attach great importance to protection against cyber risks. They have regular internal and external audits, follow procedures, and strive to keep up to date with the latest legal, operational, and technological standards for early detection of intrusions and neutralization of the attacks. Experts believe that the key to cybersecurity is people, their level of education, cybersecurity skills, and clearly delegated responsibilities. The question also remains whether to train all port staff in cybersecurity measures, or to specialize a certain number of people, who would deal exclusively with cybersecurity, or to find an optimal combination of both processes running in parallel.

The experts believe that there are many overlapping laws and regulations, while some areas of cybersecurity are under-researched and not driven by appropriate solutions. However, there should be greater harmonization of standards and legislation. There should also be closer cooperation between scientific institutes and ports, both nationally and internationally. Penalties for cyber-attacks and insurance against them are a major issue and deserve more attention in upcoming research. All this opens a whole range of cybersecurity sociological, psychological, and ethical concerns that are currently quite unresolved and should attract more research attention in the future.

Acknowledgments Research for this publication was funded by the EU Horizon2020 project 952360-MariCybERA.

Appendix

Interview Questions

Q1. You work on cybersecurity topics at the Port of Tallinn. Can you briefly tell us what your job entails?
Q2. What do you consider to be the biggest challenge in your work?
Q3. How would you rate the level of digitalization of the Port of Tallinn on a scale of 1 (lowest) to 10 (highest)? According to your experience, in what kind of correlation are the level of digitalization and the level of cyber risks? Can you elaborate on this?
Q4. Which model of cyber-security management do you adopt?—In-house, outsource, or other.
Q5. How do you check the cyber-security risk level of the port?—Through internal or external audit?
Q6. Do you report on the cyber-security risks and incidents? If yes, can you tell us how often (daily, monthly, trimester, semester) and to whom (port management, police, customs, insurance companies, banks and financial institutions, contractors and stakeholders, chamber of commerce, etc.).
Q7. How do you get informed about cyber-security risks and mitigation measures?—Thanks to security consultants (private companies), national bodies, professional bodies, chamber of commerce, industry associations of professional bodies (like IMO, EMSA, etc.), Internet, press, scientific publications?
Q8. What are your plans for closer cooperation with your partners in terms of cybersecurity?

Survey Questions

Q1. How would you rate the level of awareness of cyber risks in the Port of Tallinn? (1–5).
Q2. How would you rate the Port of Tallinn employees' level of knowledge in cybersecurity (1–5).
Q3. How much attention does the Port of Tallinn pay to cybersecurity? (1–5).
Q4. Do you expect the number of cyber-attacks in maritime to increase in the future? Y/N

Q5. In your opinion, what are the main motives for cyber-attacks?

	Money
	Political motives
	Malicious hackers' mental issues
	Combination
	Other

Q6. Kindly rank (1–3) the following cybersecurity concerns:

	Lack of regulations 1 (less important, if two other are in place)
	Lack of knowledge and understanding regarding cybersecurity issues
	Lack of the stakeholders' support in developing cybersecurity infrastructure

Q7. How likely is each of the consequences in the case of a cyber-attack in the Port? (1–5)

	Disruption of ship and/or fleet operations
	Theft of ship/cargo
	Damage to port/cargo handling equipment
	Harm to the environment
	A ground of a vessel
	Ship collision
	Physical injury or loss of life

Q8. Does the Port of Tallinn offer a well-defined cyber-security management plan? Please select the correct answer.

Yes	
No	

Q9. Which security schemes does your cyber-security plan cover? Please select the appropriate items.

ISPS (more on the physical side)
ISM (more on the physical side)
ISO 9001
ISO 17779
ISO 27001
ISO 28000
NIS2 Directive (EISS extended version; based on German standards)
Council Directive 2008/114/EC on the identification and designation of European critical infrastructures and the assessment of the need to improve their protection
SOLAS/MARPLE standards (more on the physical side)
Secured regulations (e.g., PCI Data Security Standard)
Customs and financial authorities' regulations
Other Estonian Information Security Standards (EISS) Information Technology Infrastructure Library (ITIL)

Q10. What is the port's status regarding the following security activities? If the item is covered, please tick the appropriate box.

Physical access using smart cards
Physical access using plastic-based ID cards
Physical access—automatic check-in (tickets)
Use of different authentication technologies for physical access (e.g., proximity cards, smart cards, biometrics)
Use of intrusion detection systems
Up-to-date antivirus and antispam filters
Use of firewalls
Other: Automatic—security log analytic server

Q11. Which security standards are you aware of? Check the appropriate box(es).

SIRE/ATB Inspections
ISA/IEC-62443
US Barge and Inland ATB Inspection Request
EBIS Inspection
ISO/IEC 17799:2005
ISO 20858
ISO/IEC 27001
Information Security Forum
CobIT
Information Technology Infrastructure Library (ITIL)
Capability Maturity Model Integration (CMMI)
SEISMED
None of the above
Other

References

1. G.C. Kessler, S.D. Shepard, *Maritime Cybersecurity—A Guide for Leaders and Managers* (Amazon, Great Britain, 2022)
2. A. Chiappetta, Toward cyber ports: a geopolitical and global challenge. FormaMente **12**, 95 (2017)
3. M. Afenyo, L.D. Caesar, Maritime cybersecurity threats: gaps and directions for future research. Ocean Coast. Manag. **236**, 106493 (2023). https://doi.org/10.1016/j.ocecoaman.2023.106493
4. E.P. Kechagias, G. Chatzistelios, G.A. Papadopoulos, P. Apostolou, Digital transformation of the maritime industry: a cybersecurity systemic approach. Int. J. Crit. Infrastruct. Prot. **37**(C), 100526 (2022). https://doi.org/10.1016/j.ijcip.2022.100526
5. A. Linton, Port authority role in cyber-security (2016), https://www.linkedin.com/pulse/port-authority-role-cyber-security-art-linton/. Accessed 20 Jun 2024
6. Aljazeera, Israel cyberattack caused 'total disarray' at Iran port: Report, 2020. https://www.aljazeera.com/news/2020/5/19/israel-cyberattack-caused-total-disarray-at-iran-port-report. Accessed 20 Jun 2024
7. E. Tijan, M. Jovic, S. Aksentijevic, A. Pucihar, Digital transformation in the maritime transport sector. Technol. Forecast. Soc. Chang. **170**, 120879 (2021). https://doi.org/10.1016/j.techfore.2021.12087
8. I. Peña Zarzuelo, Cybersecurity in ports and maritime industry: reasons for raising awareness on this issue. Transp. Policy **100**, 1–4 (2021). https://doi.org/10.1016/j.tranpol.2020.10.001
9. C. Senarak, Cybersecurity knowledge and skills for port facility security officers of international seaports: perspectives of IT and security personnel. Asian J. Shipp. Logist. **37**(4), 345 (2021). https://doi.org/10.1016/j.ajsl.2021.10.002
10. C. Senarak, Port security and threat: a structural model for prevention and policy development. Asian J. Shipp. Logist. **37**, 20–36 (2021). https://doi.org/10.1016/j.ajsl.2020.05.001

11. M.P. Munyai, D. Govender, Mitigation of security risks at maritime ports of entry. J. Transp. Secur. **17**(1), 11 (2024). https://doi.org/10.1007/s12198-024-00279-3
12. M.S. Karim, Maritime cybersecurity and the IMO legal instruments: sluggish response to an escalating threat? Mar. Policy **143**, 105138 (2022). https://doi.org/10.1016/j.marpol.2022.105138
13. N. Polemi, *Port Cybersecurity—Securing Critical Information, Infrastructure and Supply Chains* (Elsevier, 2018)
14. A.S. Tallinn Sadam, Consolidated group annual report 2023 (n.d.), https://www.ts.ee/en/investor/annual-reports/. Accessed 26 May 2024
15. Kalm V, The Port of Tallinn is digitalising the shipping industry, https://www.europeanceo.com/business-and-management/the-port-of-tallinn-is-digitalising-the-shipping-industry/. Accessed 26 May 2024
16. Ampron, LED electronic message boards to port of Tallinn, Estonia, [Video] (n.d.), https://ampron.eu/case_study/over-230-message-boards-to-smart-port-solution-at-port-of-tallinn/. Accessed 26 May 2024
17. Port of Tallinn, Cargo, containers (n.d.), https://www.ts.ee/en/containers/. Accessed 25 May 2024
18. HHLA TK Estonia, About us (n.d.), https://hhla-tk.ee/en/about-us. Accessed 28 May 2024
19. HHLA, Gateway to the future, Estonia makes digitalization easy, https://hhla.de/en/magazine/digital-estonia. Accessed 28 May 2024
20. Tallinn. A unique autonomous port truck is being tested at the HHLA TK Estonia terminal (2023), https://www.tallinn.ee/en/tallinnovatsioon/news/unique-autonomous-port-truck-being-tested-hhla-tk-estonia-terminal. Accessed 28 May 2024
21. L. Koper, Annual report for 2023 (2024), https://www.luka-kp.si/wp-content/uploads/2024/04/01.6-Luka-Koper_letno-porocilo_2023_kazalo.pdf. Accessed 6 Jun 2024
22. L. Koper, Cargo statistics at Port of Koper (2024), https://www.luka-kp.si/en/news/cargo-statistics/. Accessed 10 Jun 2024
23. L. Koper, Services and terminals in Port of Koper (2024), https://www.luka-kp.si/en/services-terminals/. Accessed 10 Jun 2024
24. L. Koper, *Rules on Internal Order at Luka Koper* (Luka Koper, Koper, 2011)
25. M. Bakhtina, R. Matulevičius, L. Malina, Information security and privacy management in intelligent transportation systems. Complex Syst. Informatics Model. Q. **38**, 100–131 (2024)

Open Access This chapter is licensed under the terms of the Creative Commons Attribution-NonCommercial-NoDerivatives 4.0 International License (http://creativecommons.org/licenses/by-nc-nd/4.0/), which permits any noncommercial use, sharing, distribution and reproduction in any medium or format, as long as you give appropriate credit to the original author(s) and the source, provide a link to the Creative Commons license and indicate if you modified the licensed material. You do not have permission under this license to share adapted material derived from this chapter or parts of it.

The images or other third party material in this chapter are included in the chapter's Creative Commons license, unless indicated otherwise in a credit line to the material. If material is not included in the chapter's Creative Commons license and your intended use is not permitted by statutory regulation or exceeds the permitted use, you will need to obtain permission directly from the copyright holder.

Simulating Cyber-Attacks on the Unmanned Sea-Surface Vessel's Rudder Controller

Igor Astrov and Sanja Bauk

1 Introduction

The sea is an ecosystem that is vital to life on our planet. It produces up to half of the world's oxygen supply, absorbs carbon dioxide from the atmosphere, supplies nearly three billion people with essential protein, controls the planet's temperature, and provides a variety of other resources that are used by humans. Maritime transport is the safest and the most effective in moving goods and raw materials around the world, and therefore it facilitates about 80% of world trade. However, significant pressures from climate change, ocean acidification, sea level rise, fluctuating fish stocks, natural and man-made disasters, and other factors pose a threat to the sustainability of marine habitats and coastal communities. Furthermore, a dilemma facing the maritime industry and business in the current digital era is its relative lack of digitalization when compared to other spheres of economic activity. For instance, the aviation industry, being the closest to the maritime one, has a significant advantage in terms of digitalization. Namely, there are millions of registered drones, but very few autonomous research, passenger, cargo, and military vessels [1–3].

What could be the reasons for this? Maritime is older than aviation and consequently more conservative. The regulation is not unique, and even where it is in place, it is not always adhered to. Almost all countries in the world have signed the SOLAS (Safety of Life at Sea) Convention, but there is about a third of non-SOLAS vessels like fishing boats, pleasure yachts, and small cargo vessels

---

I. Astrov
Department of Software Science, Tallinn University of Technology, Tallinn, Estonia
e-mail: igor.astrov@taltech.ee

S. Bauk (✉)
Estonian Maritime Academy, Tallinn University of Technology, Tallinn, Estonia
e-mail: sanja.bauk@taltech.ee

sailing in international sea waters. In addition, there is a lot of competition in the shipping sector, and some players are hesitant to publicly disclose information that is essential to their sustainability performance. Digitalization also brings with it the risk of cyber-attacks. Investments in digital assets are huge, while the return on revenue is vague. Therefore, the stakeholders are reluctant toward faster diffusion of emerging technologies. This could explain, to a certain extent, relatively slow pace of digitalization in shipping and port logistics on a global scale.

However, the acceleration of digitalization in the maritime sector began with the commercialization of the Internet. In the early 1990s of the twentieth century, ECDIS (Electronic Chart Display and Information System) appeared on ships as an interface to the integrated bridge system. With the digitalization of navigation, the ship's propulsion system is also becoming increasingly digital. The navigation bridge and the ship's mechanical complex are increasingly becoming like computer centers for collecting, processing, storing, and displaying data for safe, effective, and efficient navigation. In a relatively short time, there has been a diversification between IT (Information Technology) and OT (Operation Technology), and now we are moving again toward their convergence. These processes are monitored by land-based remote operation centers (ROC) and VTS (Vessel Traffic Service) centers as well as satellite CNS (Communication, Navigation, and Surveillance) systems. In ports, administrative tasks and some operational processes are being digitized and/or digitalized as physical and computer systems increasingly merge. Today, major seaports are using EDI (Electronic Data Interchange), PCS (Port Community System), SW (Single Window), and ELM (Electronic Logistics Marketplace). The Internet of Everything (IoE) appears in the second decade of the twenty-first century, further connecting the virtual and real worlds while allowing parallel micro and macro control and management [4]. Programs and data are being moved from data centers on ships, traffic control centers, and ports to the Cloud. As a result, the amount of data collected in real time is growing exponentially, opening new fields of big data and OSINT (Open-Source Intelligence). Big data and OSINT open opportunities to gain a more detailed and comprehensive insight into real-world processes in general, including in the maritime industry. Work is underway to develop underwater data collection and processing centers to make them more cost-effective and energy-efficient. Big data and OSINT will enhance the development of digital twins of ships and ports (sub-) systems, further strengthening the link between the real and virtual worlds. This is followed by the development of automated expert systems based on machine learning and artificial intelligence. The maritime digitalization corps includes surface and underwater vessels with varying degrees of autonomy. These vessels are remotely controlled, and some can operate completely autonomously for a period. Most vessels cooperate with underwater and air assets such as drones and low-orbit pseudo-satellites [5]. Digitalization in shipping and port logistics is moving in several directions, with increasing emphasis on the hazardous operation of complex systems in maritime. As digitalization increases, so does the risk of cyber-attacks. As a result, efforts are being made to detect and prevent various types of cyber-attacks in maritime.

The following parts of this chapter are structured to provide a brief overview of the ECDIS, integrated navigation bridge, e-navigation, recently developed autonomous sea-surface vessels, and autopilot concept as verticals toward maritime industry digital transformation. Finally, the cyber-attacks, on the rudder's controller of the USV "Nymo" with a known mathematical model are simulated and countered by optimizing the rudder's control system as an example of avoiding a tentative cyber-attack.

2 Electronic Chart Display and Information System (ECDIS)

Navigation has been revolutionized by ECDIS (Electronic Chart Display and Information System). An ECDIS is a system that displays hydrographic information, which is combined with information provided by electronic position-fixing systems, GNSS (Global Navigation Satellite System), radar (Radio Detection/Direction and Ranging), ARPA (Automatic Radar Plotting Aid), etc., to assist in the safe navigation of the vessel. An ECDIS consists of the Electronic Navigation Chart (ENC) as a data file, and the Electronic Chart Display Equipment (ECDE) hardware. In addition, ECDIS is a system that can store and use information from a List of lights, Sailing directions, Tide tables, etc., together with the electronic charts. An Electronic Chart System (ECS) is a generic term for equipment that displays electronic charts, but which does not satisfy the SOLAS Convention requirements. The IMO (International Maritime Organization) stated that ECDIS shall be a system that contributes to safer navigation; can replace ordinary paper charts; shows all the necessary information for safe navigation; contributes to simpler chart updating; reduces the workload on the bridge; gives the necessary alarms regarding system error; integrates navigation information such as those obtained by GPS (Global Positioning System), radar, ARPA, AIS (Automatic Identification System), gyrocompass, echo sounder, VDR (Voyage Data Recorder), and alike; and, has at least the same availability and reliability as paper charts [6–11]. More specifically, some of the key features of ECDIS are as follows:

- At any time, it shows the ship's position on the chart, in true motion with or without radar overlay or ARPA integration in one display.
- Uses the chart information necessary for safe navigation.
- Updates the chart by satellite or VHF (Very High Frequency) link and thereby increases the speed of updating.
- Allows automatic chart updating.
- Provides anti-ground warning.
- Gives tide-corrected depth contour.
- Offers storing in a "black box" the ship's navigation data for a certain period.
- Gives detailed information about lights, buoys, beacons, etc., by pointing to the chart symbol.
- Varies the chart scale according to navigation conditions.

- Varies the chart colors depending on the day or night navigation.
- Shows the ship in true size in narrow and difficult areas and conditions, etc.

Presenting navigational chart information on radar displays and or on special navigational displays is not a new idea. In the 1970s of the past twentieth century, several manufacturers offered such a system. Due to the technical limitations and high prices, even the most sophisticated of these early systems were only delivered in small numbers to ships. The *Nordcontrol Databridge* was one of the early systems providing limited chart information on the radar display. Another system with this capacity was the *Mitsubiski Tonac*. The early integrated navigation systems which had the possibility of displaying a limited amount of chart information used the Navy Navigation Satellite System (NNSS), Decca, Loran, radar positioning, astronavigation, terrestrial navigation, or Dead Reckoning (DR) to position the ship. The more advanced of these earlier integrated navigation systems consider the position information from several sources, for example, NNSS and Decca, NNSS and Loran, NNSS and DR, etc. For the professional navigators, who knew the system's limitations and always took these into account, these systems provided great assistance to navigation on many ships and certainly made navigators' jobs easier and more interesting. The early systems like today's ECDIS, if used by untrained/unqualified personnel, pose a threat to the ship and crew and other ships and the environment. Unfortunately, the new systems have not been made easier to use than the earlier ones. The complexity of today's system poses a threat and far more attention and resources must be put into Human Machine Interface (HMI) to make the system more user-friendly for the navigator. Electronic charts have been used in military applications for several decades. However, it was not until November 1988 that the International Hydrographic Organization (IHO) set up a working group to develop specifications for chart symbols and color definitions that could be evaluated by the hydrographic offices, ECDIS users, and manufacturers. Considering the many occurrences of running aground because the navigator had lost an overview of the position, the authorities started several projects to investigate the possibilities of developing a new navigational concept based on electronic charts. This was regarded as an anti-grounding system, which in coordination with the anti-collision system (radar/ARPA) was to be the "perfect" navigation system.

An ECDIS is just as reliable as the user's knowledge. The ECDIS must be connected to the ship's position-fixing system like GPS, to the gyro compass, and the speed log. If one of these sensors is not active, the ECDIS is not compliant with the IMO Performance Standard. Depending on the ship's type and age additional equipment such as a Voyage Data Recorder (VDR), Bridge Navigation Watch Alarm System (BNWAS), and Bridge Alarm Management System (BAMS) might be mandatory, while supporting systems such as DGPS (Differential Global Positioning System), radar/ARPA, autopilot, echo sounder, AIS or VDES (VHF Data Exchange System), Navtex (Navigational Telex) or NAVDAT (Navigation Data) [12], communication systems (satellite/mobile), wind sensor, slave ECDIS display, SW-key (or so-called dongle), external data files, etc., might be included. One can choose between many different designs of ECDIS.

Today the ECDIS system is often connected to an integrated bridge system or forms a part of this system, which integrates radar, ARPA, autopilot, positioning, routing, log, gyro, etc. An integrated bridge system allows these systems to work as "one system," while several options for "automatic sailing" became available to the navigator. In such an arrangement, ECDIS should not be considered as a separate functional block. Instead, chart functions are embedded in a multisensory multifunctional navigation environment which pre-selects correlated information for presentation. The appearance of integrated navigation has resulted in a more sophisticated approach to systems in terms of increased functional capabilities. However, the increased degree of integration between previously independent systems and equipment means that there are complex linkages, dependencies, and inter-relationships with an integrated navigation system.

3 E-Navigation

E-Navigation is an IMO initiative that aims to integrate existing and new shipboard and shore-based navigational tools into an "all-embracing system." It is an operational concept for marine navigation as a collaborative arrangement between onboard and land-based navigational facilities continually cooperating via broadband communication channels. The core objectives of e-Navigation approved by IMO have the following tasks:

- To facilitate safe and secure navigation of vessels having regard to hydrographic, meteorological, and navigational information and risks.
- To facilitate vessel traffic observation and management from shore/coastal facilities, where appropriate.
- To facilitate communications, including data exchange, among ship-to-ship, ship-to-shore, shore-to-ship, shore to shore, and other users.
- To provide opportunities for improving the efficiency of transport and logistics.
- To support the effective operation of contingency response, and search and rescue services.
- To demonstrate different levels of accuracy, integrity, and continuity appropriate to a safety-critical system.
- To integrate and present information onboard and ashore through a human–machine interface which maximizes navigational safety benefits and minimizes any risks of confusion or misinterpretation on the part of the user.
- To integrate and present information onboard and ashore to manage the workload of the users, while also motivating and engaging the user and supporting decision-making.
- To incorporate training and familiarization requirements for the users throughout the development and implementation processes.
- To facilitate global coverage, consistent standards, and arrangements, and mutual compatibility and interoperability of equipment, systems, symbology, and oper-

ational procedures, to avoid potential conflicts between users, and support scalability, to facilitate use by all potential maritime users.

If we go back now to the on-board decision-making, depending on the ship position, i.e., at open sea, coastal, or restricted waters, the navigator may select between several sailing options. Here are some examples of possible sailing solutions found on an integrated ship bridge system (possibly within the e-Navigation framework):

- *Course Mode*: It is a sailing mode normally used in open waters and for long-distance sailing, as this mode will give the shortest distance between two points. No correction for offset is made, but the ship will head to the destination.
- *Corrected Course Mode*: It is used in waters where it is necessary to correct for wind and current. Correction of offset is made, but no attempt to follow the original planned track is made.
- *Track Mode*: In this mode, the system will calculate the optimal path back to the planned track. Track mode is used in restricted waters whenever it is important to stay exactly on the planned route.

For a professional navigator, it's a matter of course that the route selected for actual sailing is properly checked before it is activated and used for actual sailing. Parameters used when planning the route must still be valid to maintain required safety margins, if not the route may have to be changed before it can be used safely. Examples of parameters, that may have changed after the selected route was programmed, are ship draught, position accuracy, engine reliability, steering gear reliance, etc.

Navigation with an ECDIS system, especially when the ECDIS is connected to an integrated bridge system, and e-navigation facilities, changes the work situation for the navigator a lot. A good working ECDIS reduces the navigator's workload a lot, especially when connected to a properly working integrated system, then the navigator's role is changed from actually doing the various tasks to monitoring these tasks. Seen from a safety point of view, this should be very good since it gives the navigator more time to check important parameters and monitor the traffic more closely. Sailing with the ECDIS system, as a part of an integrated bridge system, within the e-Navigation environment, requires a highly qualified navigator with a positive skepticism toward computerized systems, which can overcome serious technology overreliance issue.

4 Unmanned Sea-Surface Vessels (USVs)

The maritime industry has been implementing technological advancements related to autonomous ships since the 2010s. Autonomous vessels aim to decrease the number of ship accidents, the majority of which are believed to be caused by human factors. According to Nakashima et al. [13], whose work consider several previous studies, with autonomous vessels where the crew workload is reduced, fuel efficiency is improved, and cargo holds are expanded through the removal of the navigation bridge and crew premises. This creates a space for greater freedom in ship design. Furthermore, autonomous vessels eliminate seafarers' family separation and social isolation, along with hostage situations with pirates. Nevertheless, there are numerous obstacles in the way of putting this new technology and strategy into practice. Some of these concerns are high implementation risks, vague return of investment, economic feasibility, legal framework, COLREG (International Regulations for Preventing Collisions at Sea) for autonomous vessels, diffusion dynamics, subsidies for manufacturers and demonstrators, etc. As systems become more autonomous and complex, the risk of cyber-attacks increases. Thus, the main objective of this chapter will be to simulate a cyber-attack on the rudder controller of the autonomous ship "Nymo," along with an attempt to mitigate it.

The study [13] indicates that semi-autonomous navigation will be introduced in 2034 in Japan, while fully autonomous navigation will be introduced in 2045. Key stakeholders in this process are identified following entities:

- The producers (shipyards, producers of marine equipment, and other businesses) who create technology and carry out research and development for autonomous ships.
- The shipping firms that buy the ships and offer maintenance services.
- The decision-makers (government, classification societies, and other outside parties) who back the growth of industry and the achievement of scalable, safe maritime transportation.

These days, a few unmanned and autonomous ships are employed for commercial, military, and research objectives. Some of these are introduced below:

- The self-sufficient, short-range, zero-emission cargo ship *Yara Birkland* was designed by Yara in collaboration with Kongsberg [14]. It can hold 120 twenty equivalent unit (TEU) containers and travels along the Norwegian coast. This self-sufficient ship is being used to transport mineral fertilizer from Yara's Porsgrunn production facility to the nearby export port of Brevik. Its maximum speed is 15 knots, and its length is 80 m.
- Another autonomous vessel, the *Maxlimer* from SEA-KIT, successfully finished its maiden voyage in 2019 between the UK and Belgium. It is 11.75 m long. It was the first ship of its type to be certified by Lloyd's Register for unmanned marine systems in 2021 [15].
- Large Unmanned Surface Vehicles (LUSVs) and Extra Large Unmanned Undersea Vehicles (XLUUVs) are being developed by the US Navy to carry a variety of military payloads [16].
- The Chinese autonomous mother ship *Zhu Hai Yun*, which can launch swarms of unmanned aerial, surface, and underwater vehicles for observation and research, should also be mentioned. Zhu Hai Yun's aluminum hull has a gross tonnage of 2548 tons and dimensions of 88 m by 14 m. The vessel can reach a speed of 18 knots thanks to its diesel-electric propulsion system, which consists of two generators, two azimuth thrusters, and emergency batteries [17].

- The *Mayflower* autonomous vessel, built by IBM and MarePro, should not be disregarded. This is an experimental ship. Without a crew, it made the 40-day trip from Plymouth, UK, to Halifax, Nova Scotia, in 2022 with success. The ship is made of aluminum and composite materials; measures 49 ft long by 20 ft wide and weighs about 5 tons. After the Mayflower's 3500-mile voyage, interested parties had access to an online dashboard with a wealth of real-time data, including live video, energy consumption, speed, and weather. Even though the main goal of this autonomous ship is oceanographic research, it was built for scientific experimentation [18].
- Furthermore, the autonomous surface ship *KASS*, which is being constructed in South Korea, is still in the development stage as of 2025 [19].
- It is appropriate to mention the Rolls-Royce *AAWA* multipurpose ocean-going reduced crew ship here. This vessel is expected to be completely autonomous by 2035 [20, 21].
- Lastly, a start-up company MindChip, and the Tallinn University of Technology developed the USV known as *Nymo*. This catamaran is used for both environmental monitoring and parcel transportation. This 200 kg, 2.5-m-long vessel can carry up to 100 kg of cargo and travel 50 km at a time. In this work, simulation experiments with the rudder's control system cyber-attack intrusion and neutralization are conducted using the mathematical model of this USV and the appropriate software tools available in the MATLAB/Simulink environment.

5 Autopilot

The autopilot is a common component of all the advanced navigation systems that have been introduced so far. Early in the twentieth century, the development of the gyro compass coincided with the creation of the first automatic steering system. The system was first installed on navy ships alongside the gyro compass, and eventually, they were referred to as autopilot or gyro pilot. It spread throughout the merchant navy and fishing fleet in the 1950s and 1960s. The magnetic or gyro compass could be the source of connection for the system. The autopilot's installation had a significant impact on bridge procedures, and initially, there was a lot of doubt about the possibility of reducing safety and alertness. When the ship was operating on autopilot, poor lookout led to multiple accidents and groundings. A system failure that resulted in incorrect rudder movement also occurred. The operator had limited time to prevent deadly mishaps in narrow waters. As a result of numerous anomalies, procedures, and regulations were developed on board to guarantee the autopilot operated safely. Simple autopilots are currently widely manufactured and can be found on many recreational boats as well as the smallest fishing boats. Additionally, advancements have been made toward far more sophisticated steering systems, like adaptive autopilots and DP (Dynamic Positioning) systems. In the early 1970s, *Nordcontrol* (later renamed *Kongsberg*) pioneered waypoint tracking and integration with its sophisticated *Databridge* system. The cost of the most sophisticated DP systems can reach several million USD, while the most basic autopilots cost around 100 USD. The old distinctions between the DP and autopilot systems are also going to be eliminated because of the development [22].

Autopilots are increasingly being connected to navigation systems regularly. The conditions are ideal for integration if we have an autopilot with waypoint tracking ("track pilot") and an ECDIS (perhaps even an ECS). We often refer to these autopilots as adaptive. Normally, the autopilot will prioritize the course based on signals from the compass. The autopilot will additionally receive information from the navigation system about deviations from the route that is planned in the ECDIS if the system's track mode is enabled. "Cross-track error" is the term for this deviation, which is frequently shortened to "XTE." For instance, the autopilot will respond with a port rudder if the ship deviates too far from the starboard of the intended course, and the system will alert the user with a starboard XTE. After that, the ship will proceed along the intended course. However, it is possible that the rudder controller could be the target of a cyber-attack and the autopilot could be disrupted. In this case, it will send the wrong message to the rudder, which could result in a stroke, collision, and/or sinking of the USV (Fig. 1). Therefore, we shall carry out some simulation experiments with fictitious cyber-attacks, including mitigation of their undesired effects in the next section.

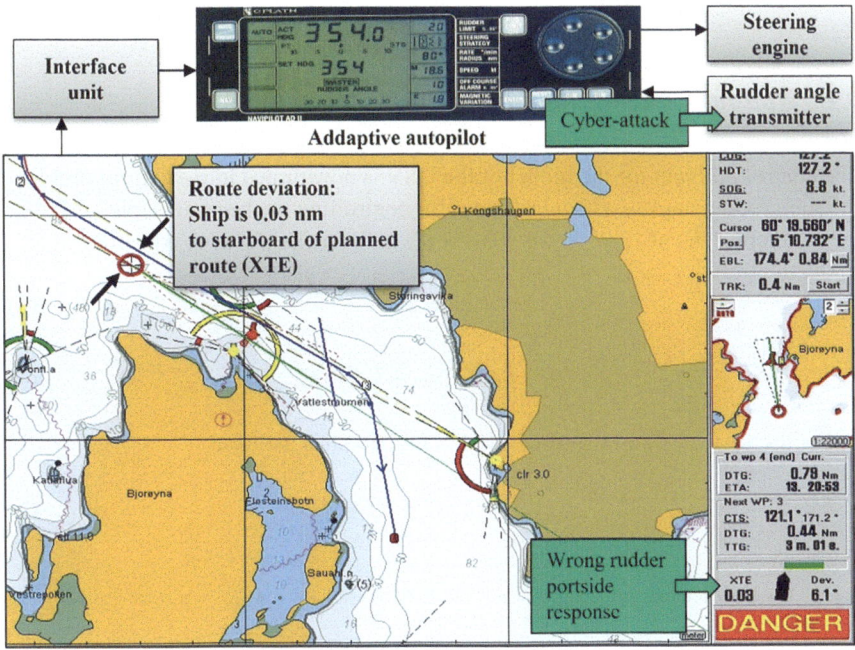

Fig. 1 Autopilot as a target of a cyber-attack. (Adopted from [22])

6 Simulation Experiment

At Tallinn University of Technology (TalTech), during a time one of the study fields has been the research of simulation models of the USVs [23–28]. Later, such an area as simulation experiments of cyber-attacks on USVs appeared [29]. Consider the model of the USV (Fig. 2). A rudder deflection causes the yaw perturbation of the USV.

The heading angle ψ can be calculated so

$$\psi(\tau) = \psi(0) + \int_0^\tau r(t)dt \qquad (1)$$

The transfer function from the input rudder angle δ to the output heading angle ψ is obtained as [30].

$$W(s) = \frac{n_2}{s(d_2 + d_1 s)} \qquad (2)$$

where

$$n_2 = 0.1; d_1 = 6.65; d_2 = 1.$$

6.1 Rudder Control System

The control problem for the model of the USV is now turned into a control problem by using rudder angle δ as control input for controlling the heading angle ψ. Using the next replacement $x_1 = \psi$, we convert (2) to the next system of equations

$$\dot{x}_1 = x_2 \qquad (3)$$

$$\dot{x}_2 = -0.15 x_2 + 0.015 \delta \qquad (4)$$

For the system (3)–(4), the Hamilton's function has the form

$$H = f_1 \varphi_1 + f_2 \varphi_2 \qquad (5)$$

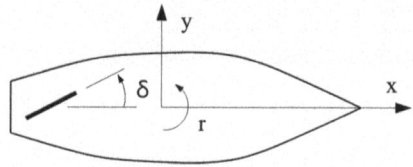

Fig. 2 The USV reference system

where

$$f_1 = x_2$$

$$f_2 = -0.15x_2 + 0.015\delta.$$

From (5), we find

$$\dot{\varphi}_1 = -\frac{\partial H}{\partial x_1} = 0 \tag{6}$$

$$\dot{\varphi}_2 = -\frac{\partial H}{\partial x_2} = -\varphi_1 + 0.15\varphi_2 \tag{7}$$

The general solution of the system (6)–(7) with respect to φ_2 can be written so

$$\varphi_2(t) = C_1 + C_2 e^{0.15t} \tag{8}$$

where C_1, C_2 are the constants. Using the general solution (8), we can determine the number of sign changes of the function φ_2. Note that the function $\varphi_2(t)$ at $C_1 = 0$, $C_2 = 1$ is an increasing curve passing through the point (0, 1) and approaching the abscissa without limit. Thus, the function φ_2, and therefore the optimal input control δ, can change sign no more than once during the transient process. Let us find the equations of the phase trajectories of the system (3)–(4). The equations (3)–(4) can be rewritten as

$$\frac{dx_1}{dt} = x_2 \tag{9}$$

$$\frac{dx_2}{dt} = -0.15x_2 + 0.015\delta \tag{10}$$

Transforming (9)–(10), we get

$$dx_1 = \frac{x_2 dx_2}{-0.15x_2 + 0.015\delta} \tag{11}$$

Having put $\delta = -1$, we rewrite (11) so

$$dx_1 = \frac{x_2 dx_2}{-0.15x_2 - 0.015} \tag{12}$$

Integrating (12), we get

$$\int_{x_{10}}^{x_1} dx_1 = x_1 - x_{10} = \int_{x_{20}}^{x_2} \frac{x_2 dx_2}{(-0.015 - 0.15 x_2)} =$$
$$= \frac{-1}{1.5}(1 + 10 x_2 - \ln(1 + 10 x_2) - 1 - 10 x_{20} + \ln(1 + 10 x_{20})) \qquad (13)$$

Putting $x_{10} = x_{20} = 0$ in (13), we get a left branch of the switching line equation of phase trajectory passing through the origin of coordinates

$$x_1 = -6.65 x_2 + 0.665 \ln(1 + 10 x_2) \qquad (14)$$

where

$$x_2 > 0, \delta = -1.$$

Similarly, the equation of the right branch of the switching line has the form

$$x_1 = -6.65 x_2 - 0.665 \ln(1 - 10 x_2) \qquad (15)$$

where

$$x_2 < 0, \delta = +1.$$

Combining the equations (14)–(15) and denoting the points belonging to the switching line with a symbol $_*$, we obtain

$$x_1^* = -6.65 x_2^* + 0.665 \operatorname{sign}(x_2^*) \ln(1 + 10 |x_2^*|) \qquad (16)$$

Therefore, optimal control is defined as follows

$$\delta = K \operatorname{sign}(\psi_d + x_1^* - x_1) \qquad (17)$$

where K is a constant, ψ_d is the desired heading angle, x_1^* is calculated by the formulas (3) and (16), x_1 is a current value of the heading angle in the control feedback branch. This equation (17) determines the control law as a function of phase coordinates. The control system configuration to regulate the input variable δ with using a relay control system is thus designed, to have the next structure (Fig. 3). The input and output signals are shown in Figs. 4 and 5, respectively. This control system allows high accuracy to achieve smooth and fast stabilization of the heading angle.

Simulating Cyber-Attacks on the Unmanned Sea-Surface Vessel's Rudder Controller 95

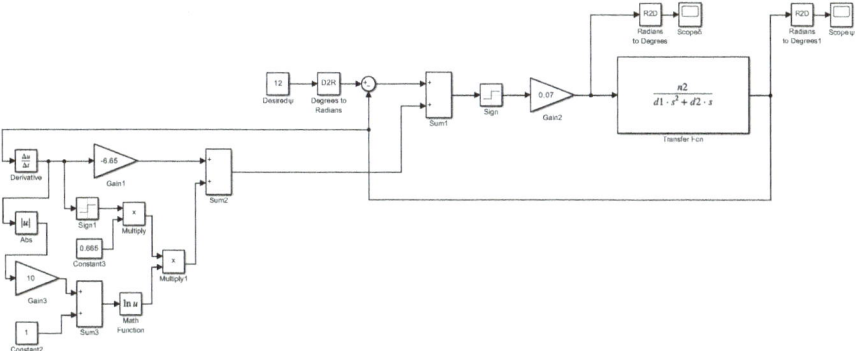

Fig. 3 Simulink-style block diagram of the USV with relay controller without disturbances

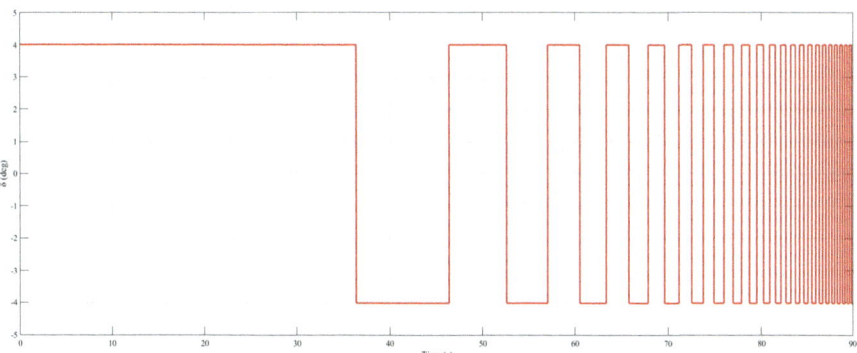

Fig. 4 Rudder angle of the USV without disturbances

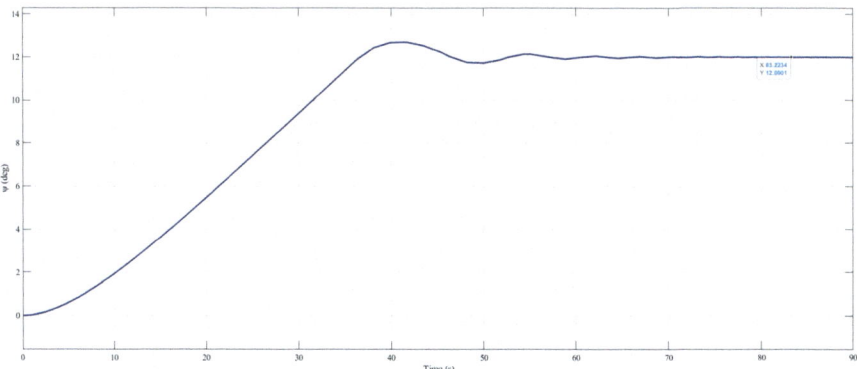

Fig. 5 Heading angle of the USV without disturbances

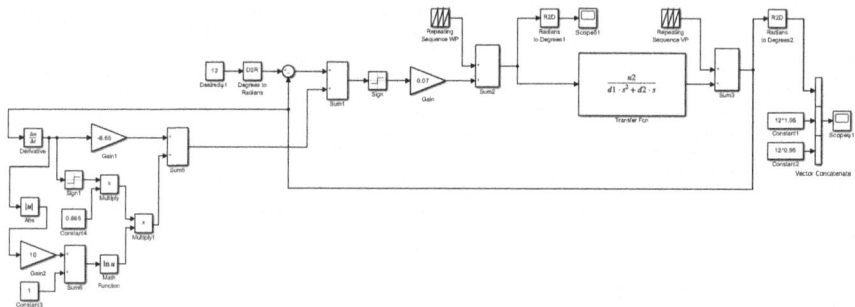

Fig. 6 Simulink-style block diagram of the USV with relay controller with disturbances

6.2 Simulation of Cyber-Attack

The adversary can attack the system on any control system component. This attack is propagated through the control system and modifies the value of the control signal in some random manner. The effect of a cyber-attack can be modeled as adding the arbitrarily shaped periodic signal having a sawtooth waveform to any control system component. The actual measurement ψ_a is related to the noiseless output ψ_n with ν is a measurement noise so (Fig. 6)

$$\psi_a(\tau) = \psi_n(\tau) + \nu(\tau) \tag{18}$$

Consider now a feedback system with a disturbance occurring at the input side of the plant as illustrated in Fig. 5.

$$\delta_a(\tau) = \delta_n(\tau) + w(\tau) \tag{19}$$

Given the model of the USV is a model with noises, it is hard to define a simple control system that follows a reference under all conditions. The control system configuration is the same as considered before, i.e., using a relay control system (see Fig. 6). The desired heading angle $\psi_d = 12$ deg, the waveform w that repeats every 1 s from the start of the simulation and has a maximum amplitude of 0.005, and the waveform that repeats every 1 s from the start of the simulation and has a maximum amplitude of 0.008 from (18) to (19) for system (2) and were applied during this maneuver. Noisy input and output signals are shown in Figs. 7 and 8, respectively. The control of the USV in such conditions becomes quasi-optimal. This control system maintains control accuracy within a 5% limit but appears to have oscillatory movements to the left and right of the chosen course.

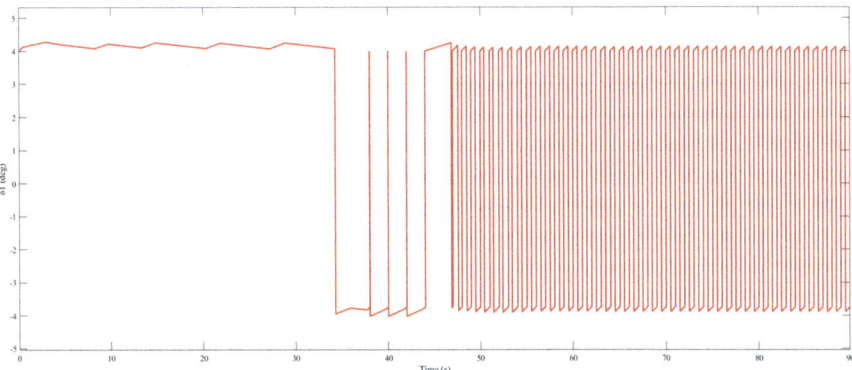

Fig. 7 Rudder angle of the USV with disturbances

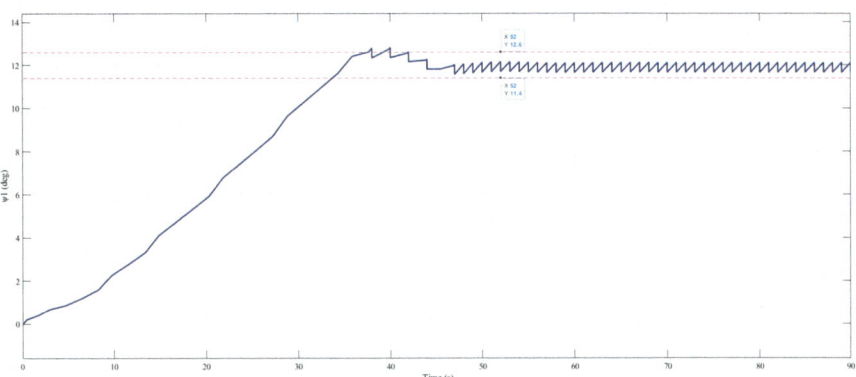

Fig. 8 Heading angle of the USV with disturbances

6.3 Simulation of Cyber-Attack Prevention

The control system configuration is the same as considered before, i.e., using a relay control system with the addition of a Kalman filter (see Fig. 9). The desired heading angle $\psi_d = 12$ deg, the waveform w that repeats every 1 s from the start of the simulation and has a maximum amplitude of 0.005, and the waveform v that repeats every 1 s from the start of the simulation and has a maximum amplitude of 0.008 from (18) to (19) for system (2) were applied during this maneuver. The Kalman filter matrix gain L is designed so that the continuous, stationary Kalman filter:

$$\dot{X}_e = AX_e + BU + L(Y - CX_e - DU) \tag{20}$$

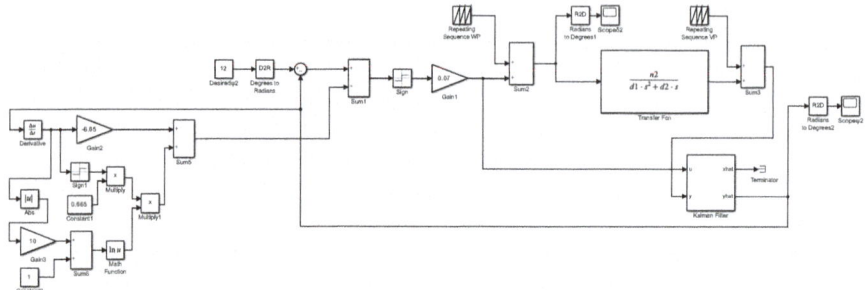

Fig. 9 Simulink-style block diagram of the USV with relay controller and Kalman filtering

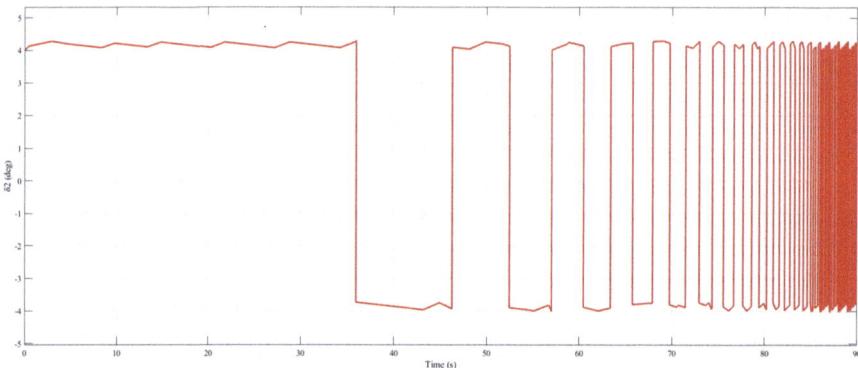

Fig. 10 Rudder angle of the USV with Kalman filtering

produces an optimal estimate X_e of vector X. The matrices A, B, C, D and vector L in (20) are calculated so

$$A = \begin{bmatrix} -0.1504 & 0 \\ 1 & 0 \end{bmatrix},$$

$$B = \begin{bmatrix} 1 \\ 0 \end{bmatrix}, C = [0\ 0.0150], D = [0],$$

$$L = \begin{bmatrix} 0.0031 \\ 0.6517 \end{bmatrix}.$$

Filtered input and output signals by the Kalman filter are presented in Figs. 10 and 11, respectively. The use of the Kalman filter demonstrates the ability to reduce to a negligible level the impact of sawtooth waveform noises caused by cyber-attacks' influence on the USV.

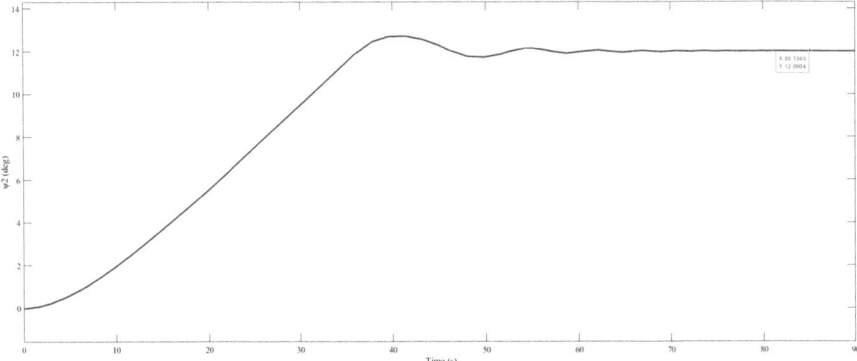

Fig. 11 Heading angle of the USV with Kalman filtering

7 Conclusion

A simulation of a cyber-attack during a maneuver of the USV's course alteration using MATLAB/Simulink was carried out. The control system for the USV is designed so that the asset can remain close to the desired course when exposed to cyber-attacks simultaneously on the input and output of the system. This control system contains a relay controller and a Kalman filter as its main elements and demonstrates high efficiency for the selected maneuver. The developed simulation model of a cyber-attack upon the mathematical model of the USV will present an important part of the development of software for the optimal control and exploitation of the actual USV. In the case of a malicious attack, a relay control system is less likely to fail because relays are highly reliable and can operate in harsh conditions. In addition to this, this chapter briefly describes key concepts and systems for the development of modern navigation, including ECDIS, integrated bridge systems, e-navigation concepts, and sea-surface autonomous vessels. This is to provide a broader perspective for the experiment involving a cyber-attack on the rudder's controller input and output signals. In the age of autonomous vessels, all ship management systems will be fully relocated to a remote-control center placed on land. This will upend the way we think about navigation and take us into uncharted territory that is full of opportunities but also unknowns and dangers, like cyber-attacks, which we have to mitigate or eliminate.

Acknowledgments Research for this publication was funded by the EU Horizon2020 project 952360-MariCybERA.

References

1. N. Kapidani, S. Bauk, I.E. Davidson, Digitalization in developing maritime business environments towards ensuring sustainability. Sustainability **12**(2), 9235 (2020). https://doi.org/10.3390/su12219235
2. S. Bauk, On maritime digitalization in emerging environments, in *Logistics Engineering*, ed. by S.J.S. Chelladurai, S. Mayilswamy, S. Gnanasekaran, R. Thirumalaisamy, (IntechOpen, London, UK, 2022). https://doi.org/10.5772/intechopen.104185
3. S. Bauk, Performances of some autonomous assets in maritime missions. TransNav **14**(4), 875–881 (2020). https://doi.org/10.12716/1001.14.04.12
4. S. Bauk, T. Dlabač, M. Škurić, Internet of Things, high resolution management and new business models, in *2018 23rd International Scientific-Professional Conference on Information Technology (IT), Zabljak, Montenegro*, (2018). https://doi.org/10.1109/SPIT.2018.8350850
5. S. Bauk, N. Kapidani, Ž. Lukšić, F. Rodrigues, L. Sousa, Autonomous marine vehicles in sea surveillance as one of the COMPASS2020 project concerns. J. Phys. Conf. Ser. **1357**, 012045 (2020). https://doi.org/10.1088/1742-6596/1357/1/012045
6. S. Bauk, *Electronic Chart Display and Information System in Brief* (CTP Printers & More Maximum Media Design, Durban, South Africa. (ISBN: 978-0-620-97267-3), 2021), p. 124
7. H. Hecht, B. Berking, M. Jonas, L. Alexander, *The Electronic Chart—Fundamentals, Functions, Data and Other Essentials—A Textbook for ECDIS Use and Training*, 3rd edn. (Geomares Publishing, Lemmer, The Netherlands, 2011). https://scholars.unh.edu/ccom/147/
8. A. Khalique, *NAVBasics, Watchkeeping & Electronic Navigation*, 3rd ed. (Seamanship International, Edinburgh, Scotland, 2016). https://www.amazon.com/NAVBasics-3-Book-Set-3rd/dp/191499308X
9. CBT ECDIS, *Seagull AS with Made with Macromedia, C-MAP Norway, and Kongsberg Norcontrol, Version A1* (Horten, Norway, 2001)
10. VShip ECDIS Company, Specific training, distance learning, https://learn.vships.com/login/index.php. Accessed 28 Sept 2019
11. Chart World, *Navigation Experts, OFS-eGlobe G2-v.2.1, ECDIS* (On-Board Familiarization System, 2018)
12. S. Bauk, A review of NAVDAT and VDES as upgrades of maritime communication systems, in *Advances in Maritime Navigation and Safety of Sea Transportation*, ed. by A. Weintrt, T. Neuman, (The Nautical Institute, Gdynia, Poland, 2019), pp. 82–87. https://doi.org/10.1201/9780429341939
13. T. Nakashima, B. Moser, K. Hiekata, Accelerated adoption of maritime autonomous vessels by simulating the interplay of stakeholder decisions and learning. Technol. Forecast. Soc. Chang. **194**, 122710 (2023). https://doi.org/10.1016/j.techfore.2023.122710
14. Yara, MV Yara Birkeland (n.d.), https://www.yara.com/news-and-media/media-library/press-kits/yara-birkeland-press-kit/. Accessed 26 Jun 2024
15. N. Skopljak, SEA-KIT one step closer to developing first type-approved USV control system (2023), https://www.offshore-energy.biz/sea-kit-one-step-closer-to-developing-first-type-approved-usv-control-system/. Accessed 26 Jun 2024
16. Congressional Research Service, Navy large unmanned surface and undersea vehicles: background and issues for Congress (2023), https://sgp.fas.org/crs/weapons/R45757.pdf. Accessed 26 Jun 2024
17. Baird Maritime, Vessel review: Zhu Hai Yun—Chinese-built drone mothership boasts autonomous sailing systems (2023), https://www.bairdmaritime.com/work-boat-world/specialised-fields/marine-research-and-training/vessel-review-zhu-hai-yun-chinese-built-drone-mothership-boasts-autonomous-sailing-systems/. Accessed 26 Jun 2024
18. J. Yellig, Myflower autonomous ship makes unmanned Atlantic crossing (2023), https://www.iotworldtoday.com/transportation-logistics/mayflower-autonomous-ship-makes-unmanned-atlantic-crossing#close-modal. Accessed on 26 Jun 2024

19. Korean Research Institute of Ships & Ocean Engineering, Korea autonomous surface ship project detail (n.d.), https://kassproject.org/en/info/projectdetail.php. Accessed 27 Jun 2024
20. Rolls-Royce. Autonomous ships—the next step (2016), https://www.rolls-royce.com/~/media/Files/R/Rolls-Royce/documents/%20customers/marine/ship-intel/rr-ship-intel-aawa-8pg.pdf. Accessed 26 Jun 2024
21. Ship Technology, Rolls-Royce reveals vision of remote and autonomous shipping through AAWA project latest findings (2016), https://www.ship-technology.com/news/newsrolls-royce-reveals-vision-of-remote-and-autonomous-shipping-through-aawa-project-latest-findings-4863708/?cf-view. Accessed 26 Jun 2024
22. N. Kjerstad, *Electronic and Acoustic Navigation Systems for Maritime Studies*, 1st edn. (NTNU Norwegian University of Science and Technology, Alesund, Norway, 2016)
23. I. Astrov, A. Udal, I. Roasto, H. Mõlder, Target tracking by neural predictive control of the autonomous surface vessel for environment monitoring and cargo transportation applications, in *Proc. 17th Biennial Baltic Electronics Conference (BEC 2020, 6–8 October 2020, Tallinn, Estonia)*, (2020), pp. 1–4. https://doi.org/10.1109/BEC49624.2020.9277115
24. I. Astrov, A. Udal, H. Mõlder, An optimal control method for an autonomous surface vessel for environment monitoring and cargo transportation applications, in *Proc. IEEE 25th International Conference ELECTRONICS (#electronicsconf2021, June 14–16, 2021, Palanga, Lithuania)*, (2021), pp. 1–6. https://doi.org/10.1109/IEEECONF52705.2021.9467483
25. I. Astrov, A. Udal, H. Mõlder, T. Jalakas, T. Möller, Wind force model and adaptive control of catamaran model sailboat, in *Proc. 8th International Conference on Automation, Robotics and Applications (ICARA 2022, 18–20 February, 2022, Prague, Czech Republic)*, (2022), pp. 202–208. https://doi.org/10.1109/ICARA55094.2022.9738524
26. I. Astrov, A model-based PID control of turning maneuver for catamaran autonomous surface vessel, in *Proc. IEEE 9th International Conference on Engineering and Emerging Technologies (ICEET 2023, 27–28 October, 2023, Istanbul, Turkey)*, (2023), pp. 1–6. https://doi.org/10.1109/ICEET60227.2023.10526157
27. I. Astrov, I. Astrova, A model-based adaptive control of turning maneuver for catamaran autonomous surface vessel. WSEAS Trans. Syst. Control **19**, 135–142 (2024). https://doi.org/10.37394/23203.2024.19.14
28. A. Udal, J. Kaugerand, H. Mõlder, I. Astrov, S. Bauk, Modeling the trajectory tracking accuracy of an autonomous catamaran patrol vessel under different positional data disturbance conditions, in *MT'24: 10th International Conference on Maritime Transport, Barcelona, Spain, June 5–7*, (Universitat Politècnica de Catalunya, 2024), pp. 1–15. https://doi.org/10.5821/mt.13165
29. I. Astrov, S. Bauk, Simulating a cyber-attack on an autonomous sea surface vessel's rudder controller, in *Proc. 13th Mediterranean Conference on Embedded Computing (MECO 2024, June 11–14, 2024, Budva, Montenegro)*, (2024), pp. 558–564. https://doi.org/10.1109/MECO62516.2024.10577872
30. T.I. Fossen, *Guidance and Control of Ocean Vehicles* (Wiley, New York, 1994)

Open Access This chapter is licensed under the terms of the Creative Commons Attribution-NonCommercial-NoDerivatives 4.0 International License (http://creativecommons.org/licenses/by-nc-nd/4.0/), which permits any noncommercial use, sharing, distribution and reproduction in any medium or format, as long as you give appropriate credit to the original author(s) and the source, provide a link to the Creative Commons license and indicate if you modified the licensed material. You do not have permission under this license to share adapted material derived from this chapter or parts of it.

The images or other third party material in this chapter are included in the chapter's Creative Commons license, unless indicated otherwise in a credit line to the material. If material is not included in the chapter's Creative Commons license and your intended use is not permitted by statutory regulation or exceeds the permitted use, you will need to obtain permission directly from the copyright holder.

A Scope Review of Secure Broadcasting Protocols for the Automatic Identification System

Leonidas Tsiopoulos and Risto Vaarandi

1 Introduction

Shipping and the maritime infrastructure have a critical role in modern societies from the economic perspective [11]. Maritime systems have become heavily digitalised and interconnected [9, 33]. Implementation of IT technologies in maritime systems has introduced cybersecurity issues similar to the ones that can be found in traditional IT systems.

Cyberattacks can interrupt normal operations for longer periods, thus resulting in significant financial damage and potentially leading to catastrophic consequences. For example, in 2017, the Maersk logistics company network was cyberattacked resulting in substantial financial losses, estimated at 300 million US dollars [8, 10]. This case is one of many conducted against maritime infrastructures, as reported by Afenyo and Caesar [1]. Cyberattacks directly affecting the control of ships or monitoring of ship traffic have also been recently reported. For example, in 2016, a cyberattack misdirected two navy vessels in the Persian Gulf [40]. In a case from 2017, cybercriminals gained access to the navigation systems of a container vessel owned by a German company, which had a capacity of 8250 TEUs [21]. Moreover, an Italian base station of Automatic Identification System (AIS) experienced a ship-spoofing incident near Elba Island, where thousands of fake ships suddenly appeared and affected vastly the accurate monitoring of the maritime traffic in the area [3].

In response to the need to improve the cybersecurity in the maritime domain, leading maritime organisations such as the International Maritime Organization (IMO) and Baltic and International Maritime Council (BIMCO) released recommendations and guidelines on cybersecurity and maritime cyber-risk management

---

L. Tsiopoulos (✉) · R. Vaarandi
Department of Software Science, Tallinn University of Technology, Tallinn, Estonia
e-mail: leonidas.tsiopoulos@taltech.ee; risto.vaarandi@taltech.ee

© The Author(s) 2025
S. Bauk (ed.), *Maritime Cybersecurity*, Signals and Communication Technology,
https://doi.org/10.1007/978-3-031-87290-7_6

[6, 26, 28]. International Ship and Port Facility Security (ISPS) Code established by the IMO partly regulates the cybersecurity risk assessment [52]. These guidelines and regulations do not specify particular characteristics of cybersecurity solutions on a ship.

One of the key components to assure situation awareness at sea aiding safety is the AIS [2, 23]. AIS was created to offer identification and positioning data between vessels and from vessels to coastal stations, enabling them to monitor, identify, and share information about marine traffic. Vessels equipped with AIS continuously transmit their unique identification along with other vital navigation data. Despite the criticality of AIS, its communication protocol is not secured from cyberattacks. This, in turn, makes AIS a desirable target of cybercrime by altering or falsifying data and allowing the spread of inaccurate information about vessels, as exemplified by the above-mentioned real incidents [3, 40]. The two fundamental flaws of the AIS protocol are the lack of digital signatures for integrity and authenticity and the lack of encryption for confidentiality and privacy.

The absence of security for the AIS broadcast communication protocol and various ways to attack it were first demonstrated by Balduzzi, Pasta and Wilhoit, in 2014 [5], and more recently by more works such as the works by Levy et al. [37] and by Khandker et al. [35].

According to Balduzzi, Pasta and Wilhoit [5], attacks against AIS protocol can be divided into three classes—spoofing, hijacking, and availability disruption. Spoofing involves creating AIS messages with false information, for example, transmitting false data about the location of ships and navigational aids, about weather, etc. Hijacking involves modifying legitimate AIS messages, whereas availability disruption attacks are conducted over radio frequency and attempt to prevent legitimate AIS devices from transmitting.

The cybersecurity vulnerability analysis of the AIS broadcast protocol formed the base for the proposition of several possible solutions to secure AIS broadcasts that can be found in the literature up to date. Interestingly, based on the criticality of the topic, to the best of the authors' knowledge, a paper reviewing all these AIS secure protocol solutions has not yet been published. Thus, the main contribution of this chapter is the review and analysis of all published security-enhancing AIS communication protocol solutions and a discussion of possible ways forward. The literature search was conducted in Google Scholar and Web of Science using relevant keywords and by applying snowballing to identified papers.

The rest of this chapter is organised as follows. In Sect. 2 we provide the preliminaries for the AIS. In Sect. 3 we review the proposed solutions of all the selected papers categorising them into backward compatible and non-backward compatible according to the AIS protocol standard. In the same section, we also dedicate a subsection to solutions that can be considered as special cases. In Sect. 4 we discuss the main obstacles for the feasibility of the proposed solutions in practice and propose a possible way forward. In Sect. 5 we conclude the chapter.

2 Preliminaries for Automatic Identification System

The AIS is a coastal tracking system designed for short-range monitoring, typically effective up to a distance of 20–100 nautical miles (NM) at sea, depending on the setup [14, 16]. It was created to offer identification and positioning data to vessels and coastal stations, enabling them to monitor, identify, and share information about marine traffic. The IMO, under the Safety of Life at Sea (SOLAS) Convention, mandates that AIS transponders be installed on international voyaging ships with a gross tonnage (GT) of 300 or more, all ships with a GT of 500 or above, and on all passenger ships irrespective of size. This requirement aims to enhance safety and navigation efficiency at sea [27].

AIS transponders are available in two different classes. Class A transponders are required for all SOLAS-compliant vessels, as previously mentioned. Class B transponders, on the other hand, are designed for non-SOLAS vessels, including domestic commercial vessels and pleasure crafts.

The installed AIS transponder regularly broadcasts information of the ship's status, such as static details, dynamic (e.g., vessel position, speed, navigational status) data, and voyage (destination port and the estimated time of arrival of the vessel) information [25]. The dynamic AIS data is automatically transmitted every 2–10 seconds, depending on the vessel's speed as detailed in Table 1. The static data is transmitted every 6 minutes regardless of the vessel's movement speed or status [12].

Figure 1 displays the types of AIS messages defined by the AIS protocol standard [30]. Some message types have been allocated for mobile or base stations only, while some types are used for specific purposes, e.g., for aids-to-navigation (AtoN) messages, for safety messages, or for timing. For example, message type 10 is used for requesting current time from a remote AIS station, and message type 11 carries the response from the remote station. It should be pointed out that some message types like 6 and 8 have been designed for carrying application-specific binary data and that the AIS protocol standard does not define the exact nature of these binary data [30].

AIS transponders use the Time-Division Multiple Access (TDMA) protocol, which allows several consumers to use the same frequency channel, so that each consumer transmits within its own time slot. The size of AIS time slot is

Table 1 Default timing of AIS messages [25]

Type of ship	Reporting interval
Ship at anchor	180 sec
Speed 0–14 knots	12 sec
Speed 0–14 knots and changing course	4 sec
Speed 14–23 knots	6 sec
Speed 14–23 knots and changing course	2 sec
Speed >23 knots	3 sec
Speed >23 knots and changing course	2 sec

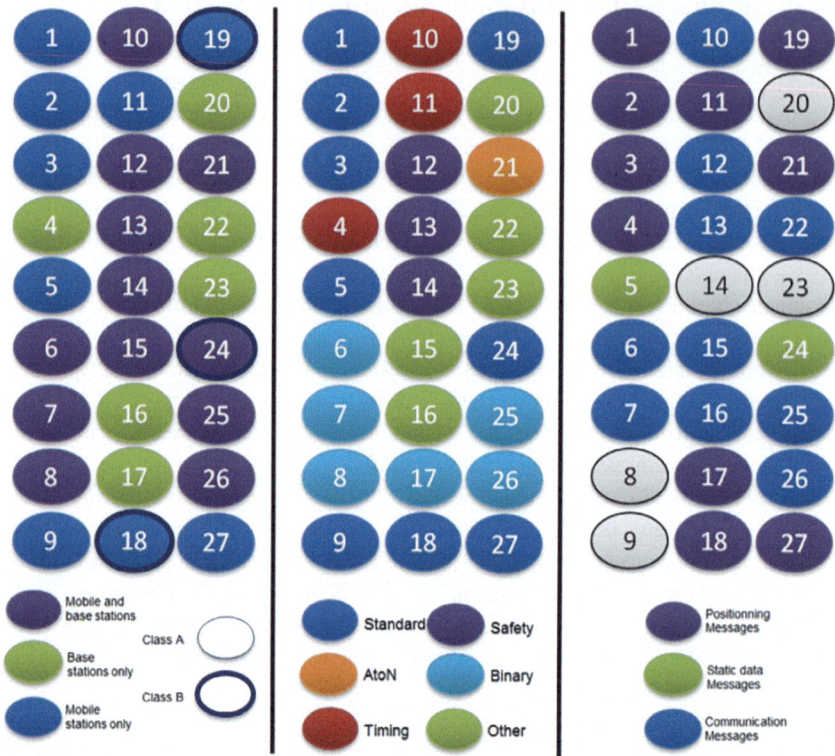

Fig. 1 AIS message types (reproduced from the work by Iphar et al. [29])

1 CHANNEL TIME SLOT – TX in 26.67 ms							
256 bits DATA PACKET							
8 bits	24 bits	8 bits	168 bits		16 bits	8 bits	24 bits
Ramp Up	Preamble	Start Flag	AIS Data Message		FCS	End Flag	Buffer

6 bits	2 bits	30 bits	2 bits	10 bits	6 bits	80 bits	32 bits
MSG ID	REPEAT INDICATOR	SOURCE ID	SPARE	DESIGNATED AREA CODE	FUNCTIONAL ID	APPLICATION DATA	RESERVED FOR BIT STUFFING

Fig. 2 An example AIS packet containing an AIS message of type 8 (reproduced from the work by Sciancalepore et al. [46])

26.67 milliseconds, and since data transmission rate is 9600 bits/sec, each slot accommodates 256 bits ($9600 \times 0.02667 \approx 256$).

Figure 2 displays an example AIS packet containing an AIS binary message of type 8. The packet begins with the 8-bit ramp up (start of transmission), followed by the 24-bit preamble (bits 0101...01) and the 8-bit start flag (bits 01111110). After the AIS message, the packet continues with the FCS (frame check sequence) field (a 16-bit CRC for error checking purposes) and the 8-bit end flag (bits 01111110). The 24-bit buffer is used for bit stuffing, synchronisation jitter, and distance delay. The example from Fig. 2 displays an AIS packet contained in a single time slot. According to AIS protocol standard [30], a sender can occupy at most five consecutive time slots for one continuous transmission. In the case of such long packets, transmission overhead (ramp up, preamble, start and stop flag, FCS, and buffer) must be used only once.

The AIS message inside the packet begins with the 6-bit message ID field (set to 8 in the example from Fig. 2 to reflect the message type), followed by the 2-bit repeat indicator (used by repeaters to indicate how many times the message has been repeated, with 0 being the default and 3 indicating "do not repeat any more"). The 30-bit source ID holds the Maritime Mobile Service Identity (MMSI) which identifies the sender of the message, while two spare bits are reserved for future use. The 10-bit designated area code (DAC) contains a maritime identification code of a country (e.g., 366 corresponds to the USA), while the 6-bit functional ID is reserved for use by the authorities inside the designated area. Finally, the AIS message from Fig. 2 contains 80 bits of application data, whereas 32 bits are used for bit stuffing. As discussed above, message type 8 has been designed for carrying application-specific binary data, and by the AIS protocol standard, the recipient of the message can determine the type of the application from the combination of the DAC and functional ID fields [30]. That convention will allow the authorities of each designated area to define their own custom use for type 8 messages (for example, US Coast Guard employs type 8 messages as a part of an AIS encrypted secure communication scheme [53]).

3 AIS Secure Protocol Solutions

In this section, we review the published papers proposing security enhancement to AIS broadcasts. In the following, we categorise the papers in different subsections depending on if the solutions are backward compatible or not to the AIS protocol standard and existing AIS equipment. Furthermore, we categorise the backward-compatible solutions to solutions that utilise alternative channels for the broadcasting of security-related information and to solutions that use AIS binary message type 8 with higher payload for the propagation of security-related information. Additionally, we dedicate a subsection to solutions that can be considered as special cases.

3.1 Backward-Compatible Solutions

3.1.1 Utilising Alternative Channels for AIS Security

Wimpenny et al. described an authentication scheme using Public Key Cryptography (PKC) allowing communications to be authenticated by the inclusion of digital signatures [55], specifically Elliptic Curve Digital Signatures (ECDS). Using a PKC scheme to authenticate communications has a considerable overhead. The difficulty lies in incorporating adequately large digital signatures into low-bandwidth VHF communications such as AIS and VHF Data Exchange System (VDES) to provide an adequate level of security. The main goal of VDES is the better exploitation of the VHF data link, and it is a proposal by IALA e-NAV Committee [24], which is also discussed at ITU, IMO, and other relevant Bodies. The intention with VDES is to use the free VHF bandwidth to facilitate eNavigation solutions such as the collection, integration, exchange, presentation, and analysis of marine information on board and ashore, to enhance navigation and related services. Considering these constraints and ongoing development of VDES, Wimpenny et al. identified a compromise and recommended the use of Elliptic Curve cryptographic algorithms, using 256-bit key size to produce signatures of 512 bits in size. Such signatures can be included in consecutive multi-slot binary AIS messages. From that work a number of open questions formed, such as the impact on the AIS/VDES channel loading and the management of the public keys needed to verify digital signatures.

Wimpenny et al. [54] described a series of improvements to the PKC authentication system for AIS described in [55] and investigated the impact of AIS channel loading due to the use of PKC. The authors discussed the problem of high overhead, since an ordinary AIS message requiring only one time slot would require four additional slots for consecutive binary AIS messages of type 8 if signed, and suggested signing only high-risk messages or batches of messages. Since the approach uses signatures of a length of 512 bits (+64 bits for a timestamp to correctly relate an AIS data message with a signature message), utilising a lot of any potentially available space to carry the payload within the existing AIS standard, an implementation was set up for VDES Application-Specific Messages (ASMs) which can accommodate a maximum payload of 1056 bits in three consecutive slots, and thus the signature message can be handled with a 2-slot ASM. The implementation consisted of using two software-defined radio systems to transmit and receive ASM messages containing the PKC signatures. Thus, an additional side channel is required, together with relevant hardware, to implement such an approach increasing the complexity and hampering feasibility.

Shyshkin proposed a method for ensuring the authentication and integrity of AIS messages based on the use of the Message Authentication Code (MAC) scheme enhanced with timestamp data and digital watermarking (WM) technology to organise an additional tag transmission channel [47]. Although the MAC length is typically 32 bits, the author justified that it is enough to detect an attack by repeatedly transmitting a very compressed version of the MAC, taking into account that AIS messages are transmitted regularly at intervals from 2 seconds to 3 minutes, thus avoiding to overload each AIS message with a full MAC. Similarly to several other works reviewed in this chapter, the proposed AIS protection employs a symmetric encryption scheme that uses the same key to generate and verify the MAC, requiring a trusted third party (TTP) for distributing the keys. To transmit the bits of the compressed MAC, the author proposed a watermarking algorithm to embed the bits into the GMSK low-frequency signal within a TDMA slot. The proposed method is fully compatible with the existing AIS functionality.

Pilosu et al. [42] explored the potential role of white-space frequencies as an auxiliary channel, bearing out-of-band signals required to make AIS more secure. Such an auxiliary channel is a complementary channel set in the TV White-Space frequencies. The concept was based on a reference architecture of an AIS class-A equipment, with a cable connected to an additional communication module sharing AIS data and sending them on an alternative secure broadband channel. The adopted communication protocols were selected from within the Wi-Fi (IEEE 802.11) standards set. At the time of this work, the VHF and UHF frequencies freed by the switch-off of analogue television (54–790 MHz) were not yet assigned to a specific service, and there was an interest in them, especially for the 700 MHz UHF band being promising for its propagation characteristics. Through extensive simulations for various traffic and obstacle scenarios, the authors investigated the applicability of the TV White-Space band to the maritime environment. The effect of the propagation conditions such as relative positions of the communicating nodes and obstructions on signal reception was pointed out. Based on the positive results on the effectiveness of maritime communications in the TV White-Space band, the authors proposed this solution as an alternative channel to work in parallel with the current AIS system for facilitating its security. A solution for the actual security protocol was not presented by the authors.

Pilosu et al. [43] in a follow-up work presented a prototype of the solution motivated in [42]. The so-called Robust Communication Module (RCM) was integrated with a commercial AIS module. The approach for the prototype implementation consisted in adding a complementary wireless communication layer in the UHF band. The RCM and AIS modules were connected via an Ethernet cable link. The AIS module sent a copy of its messages to the RCM, while the RCM shared these messages with other peers, through an encrypted and authenticated channel in the 700 MHz band, taking advantage of the Wi-Fi channel between RCMs and its built-in security features. Although the approach is promising and backward compatible, it is hard to take into use in a broader maritime context due to the requirement for additional hardware and software for AIS peer. To the best of our knowledge, this solution has not been adopted in practice.

3.1.2 AIS Binary Type 8 Message Utilisation

Goudosis, Kostis and Nikitakos proposed a solution based on the creation of a global, x.509-like Maritime Public Key Infrastructure (PKI), where the registration and Certification Authorities (CAs) would be the IMO and the national maritime authorities [19]. In this approach, the national maritime CAs would be cross-trusted or root-trusted via the IMO root certificate. Each vessel and each high-ranking marine officer would obtain a public key certificate from the maritime CA. The proposed Maritime PKI would assume responsibility for the secure distribution of the keys of symmetric algorithms when a need for encrypted data transactions between vessel-to-vessel or vessel-to-shore base stations would arise. This proposal suffers from implementation difficulties, because implementing a PKI infrastructure in a global maritime environment may prove to be a very difficult task, and because certificates are very resource demanding, incurring substantial overhead in AIS message broadcasts.

Goudossis and Katsikas [20] envisaged a potential PKI that would be run by the IMO. The authors explored how Identity-Based Public Cryptography and Symmetric Cryptography may enhance the security properties of the AIS. The concept was founded on the assumption that a Maritime Certificate-less Identity-Based Public Key Cryptography infrastructure exists. The authors motivated their proposition in that Identity-Based Cryptography (IBC) is a branch of PKC whose main characteristic feature is that the public key is a publicly known distinctive identity of each member (e.g., MMSI) needing neither CAs nor certificates. This property provides the opportunity to design simpler and less resource-demanding public key implementations. The critical properties of anonymising a vessel's AIS data using a pseudo-MMSIs implementation was then researched and a procedure was proposed. The authors also proposed the ad hoc use of encrypted AIS messages in dangerous sea areas under a TTP coordination, incorporating symmetric key cryptography, and by using the assumed infrastructure for the control of the symmetric key. This solution is similar to the commercial one by Saab which provides proprietary encrypted AIS only in vessels equipped with the same AIS product (see Sect. 3.3). Another similar approach is the Encrypted Automatic Identification System (EAIS) proposed by the United States Coast Guard [53].

In a subsequent paper, to implement the solution proposed in [20], Goudosis and Katsikas [17] incorporated the Sakai-Kasahara identity-based authentication and encryption scheme that follows the IEEE 1363.3-2013 standard for IBC. Similarly to several other works reviewed in this chapter, the authors proposed to incorporate AIS messages type 6 and 8 that permit the encapsulation of data of non-AIS dependent, arbitrary applications for the transmission of the security-related signed/encrypted data. For this, the authors introduced an application to be used as an interface to implement the scheme over the currently available AIS infrastructure, targeting transparency and backward compatibility to the original AIS protocol.

Goudosis and Katsikas [18] further built on [17, 20] and provided a proof-of-concept implementation utilising the RFC6507 for Elliptic Curve-Based Certificateless Signatures (ECCSI), and the RFC6508 for Sakai-Kasahara Key Encryption (SAKKE) standards. RFC6508 uses IBC to, e.g., securely provide the key of a symmetric cipher to a receiver, while RFC6507 provides authentication via certificateless signatures that are based on the Elliptic Curve Digital Signature Algorithm (ECDSA). The RFC6508 utilises a variant of the Sakai-Kasahara key encapsulation mechanism, optimised to support multiparty communications that have been adopted by the IEEE 1363.3 2013 standard for IBC. The RFC6507 standard is compatible with the IBC proposed in RFC6508 and is optimised to have low-bandwidth and low-computational requirements. Overall, the work by Goudosis and Katsikas is based on too strong assumptions and requirements regarding the needed maritime infrastructure to implement such approach. The idea of also encrypting the AIS communication in addition to signing it, especially in insecure areas to prohibit adversaries from getting information about vessels in the area, is problematic regarding compatibility to AIS devices of participating vessels (requiring all participating vessels to have the proposed application installed) and also requires all participating vessels to register with the IMO. Thus, such approach is infeasible to use in high-traffic areas where all vessels should be able to exchange AIS data, and would work only for small groups of trusted users, while also suffering from message overhead requirements to transmit the security-related data, which can take up to three slots using AIS message type 8.

In a paper by Struck and Stoppe [50], several backward-compatible signature schemes for AIS messages are reviewed and compared concerning their applicability. Namely, utilising spare bits in AIS position messages to incorporate the signature bits, utilising AIS binary messages to send the signature, modifying the existing architecture of transponders to extend the low modulated signal (Gaussian Minimum Shift Keying is a form of digital modulation widely used in mobile communication and for the AIS) in order to package additional data in it, and utilising alternative channels to transmit the signature as done by the works described in Sect. 3.1.1. The authors then proposed a solution based on asymmetric signatures using pairing-based elliptic curve cryptography. The signature length is 160 bits enabling its transmission within a two-slot AIS type 8 binary message and providing a level of security which is equivalent to RSA-960. Besides proving AIS authenticity, the system is backward compatible in order to avoid the replacement of every AIS transceiver.

Aziz et al. proposed *SecureAIS*, a key agreement scheme that allows any pair of vessels in the range of an AIS radio to agree on a shared session key of the desired length to be used for subsequent communications [4]. The authors integrated elliptic-curve-based certification and key agreement schemes to provide a low-overhead solution. Notably, this approach allows for secure communication between two vessels only. Thus, the broadcasting feature of AIS is not preserved.

Sciancalepore et al. [46] proposed a technique using the Timed Efficient Stream Loss-tolerant Authentication (TESLA) algorithm. The purpose of TESLA is to provide an efficient source authentication protocol for broadcast communications in resource-limited environments, by employing only symmetric cryptographic primitives. TESLA requires: (i) a loose time synchronisation between all the entities in a network, (ii) the division of the time into slots, and (iii) the deployment of a TTP to assign and manage symmetric cryptographic keys assigned uniquely to each entity in the network. The authors combined TESLA with the Bloom filter to provide two modes of authentication depending on traffic scenarios and security requirements, the "Deterministic Security Configuration" and the "Probabilistic

Security Configuration". Although one of the objectives for the technique proposed by Sciancalepore et al. was to reduce data overheads by broadcasting a signature message (utilising the AIS binary message type 8) relating to the previous N data messages, the deterministic configuration incurred a message overhead of 75%, and the probabilistic configuration incurred a message overhead of 35%. Additionally, this overhead reduction technique introduces a delay before data messages can be verified, and it can be deemed less suited for use with infrequent AIS broadcasts. On the other hand, it can be argued that this incurred delay before the information of N AIS messages can be observed might hamper situational awareness in critical cases. Furthermore, Sciancalepore et al. allow for a range of bits to be allocated to security, with the minimum being 80-bit security equivalent key sizes and the maximum 192 bits, depending on the trade-off between security requirements and message overhead. 80-bit key sizes are not considered adequately secure by NIST [7]. Last but not least, the authors did not discuss how to mitigate the TESLA security vulnerability due to loose timing synchronisation between GPS devices found by Jahanian, Amin and Jahangir when they analysed the TESLA protocol using Timed Coloured Petri Nets [31].

Saleem et al. [45] emphasised the similarities (both in the broadcast protocol and in the security vulnerabilities) between ADS-B (Automatic Dependent Surveillance-Broadcast) from the aviation domain and AIS and motivated that solutions should be uniform and interoperable across these technologies, avoiding customised stack for each technology. The authors then proposed a backward-compatible software-only solution to provide stronger security to ADS-B and AIS protocols based on a single generic PKI-enabled message-integrity and authentication scheme that can work seamlessly for the ADS-B and AIS, with the possibility of easy extension and integration into other protocols such as ACARS (Aircraft Communication Addressing and Reporting System). Focusing on the AIS protocol, the authors used ECDS for source authentication and data integrity of the AIS message. Since the message field of an AIS message is 168 bits, signed messages cannot be accommodated in this field. To overcome this problem, similarly to several other works in this subsection, the authors used AIS binary message type 8 for four consecutive slots to transmit a signature taking 512 bits with a timestamp of 64 bits. At first, an AIS data message is transmitted, and then a follow-on message with a digital signature inside message type 8 is transmitted. To link an already received AIS message with the received signed message and prevent a replay attack, hash digests with timestamps are computed. The authors assumed that PKI services are already established and readily available, and key management is available with the upgraded system. Furthermore, the systems to be upgraded with the proposed solutions are supposed to have cryptographic computation capabilities for signature generation and signature verification for both the transmitter and the receiver. The above assumptions and requirements, together with the high additional overhead, hamper the implementation of this approach.

3.2 Non-backward-Compatible Solutions

Litts et al. [38] leveraged the TESLA protocol to propose an authentication protocol that enhances the security of the AIS by providing it with a message-integrity and broadcast-authentication feature. The proposed protocol uses time-delayed key disclosures of hash chains as part of a message authentication code appended to packets to verify the authenticity of a sender. The authors proposed an update to the existing AIS standard implementing new message types which contain a modified data field that allows the creation of an asymmetric cryptography scheme. The scheme requires only one additional time slot for broadcast authentication compared to 3–4 additional slots required by backward-compatible solutions. This translates to substantially less overhead. On the other hand, the requirement to update the AIS protocol and the requirement for the participation of a central authority, such as IMO, to provide and distribute private keys, are obstacles for practical deployment. Also, the authors did not discuss how to mitigate the TESLA security vulnerability due to loose timing synchronisation between GPS devices found by Jahanian et al. [31].

Stewart et al. [48] proposed an AIS extension to ensure the integrity of AIS messages by digitally signing them, allowing to mitigate attacks which involve the modification of legitimate AIS messages. The proposed AIS extension assumes the presence of a public key infrastructure, with each vessel having a private key. For digitally signing an AIS message, the authors suggested two approaches that involve new AIS message types which carry digital signature information. With the first approach, a regular AIS message is followed by the second message of type 28 which holds the hash of the first message encrypted with a private key of the sender (the AIS protocol standard [30] defines types 1–27 and type 28 is currently unused). To verify the authenticity of the first message, the recipient uses the public key of the sender for decrypting the hash in the second message, comparing it with the hash of the first message. With the second approach, the original message and its encrypted hash are encapsulated into an AIS message of type 29 (currently unused by the AIS protocol standard). From the proposed approaches, the first retains compatibility with existing AIS equipment, since AIS receivers that do not recognise messages of type 28 will discard them, whereas original messages will still be processed as usual. However, one drawback of the first approach is the twofold increase of the AIS message volume. According to the authors, the second major drawback is the complexity of implementing a PKI with a secure private key management in the maritime domain.

Hall et al. [22] proposes a modification of AIS protocol which involves dividing AIS traffic between three tiers to enhance privacy. Tier1 messages involve broadcasting navigational information as with traditional AIS, but Tier1 messages would not contain any information that would reveal the identity of the vessels. For that purpose, the authors propose to utilise pseudonyms for vessel names as described by the IEEE 1609 standard. Tier2 messages involve encrypted queries from other vessels to reveal more information about the current vessel, so that the recipient of the query can either accept or deny the query. Finally, Tier3 messages involve queries from authorised parties (such as law enforcement agencies) which require an answer from the vessel. The approach proposed by the authors assumes setting up private keys and certificates on vessels and implementing three-tier messaging in AIS equipment, and implementing these changes on a larger scale is a long process.

Su et al. [51] proposed an improvement to AIS protocol which involves shore-based AIS centre acting as a trusted authority (TA). TA issues certificates to vessels, and vessels have to store encryption keys in the AIS equipment. The proposed method involves digitally signing AIS messages, so that recipients can verify the integrity of the messages with the sender certificate. In addition, the method proposes to hide the identity of a vessel with k pseudonyms that change over time. In order to hide the identity of the vessel from TA when certificates for pseudonyms are created, blind signatures are used. In their study [51], the authors did not discuss to what extent the proposed method would be compatible with the AIS protocol standard. To implement all the features of the method, modifications to existing AIS equipment and the creation of TAs are necessary, complicating the adoption of the method in the maritime industry.

Jegadeesan et al. [32] proposed a privacy-preserving improvement to AIS which involves maritime traffic controller (MTC), base stations, and vessels. MTC is considered a fully trusted entity, with one of its tasks including secret key generation and secure distribution to registered system entities. For preserving the privacy of vessels, they are authenticated by remote peers without knowing their true identity. In addition, the proposed scheme is able to preserve the privacy of the vessel trajectory. When compared to other approaches, the scheme requires less energy for computations and communication with remote peers [32]. The compatibility of the scheme with the AIS protocol standard was not discussed in [32], and as with previously described approaches, adopting this scheme on a larger scale is complicated by the necessity of modifying existing AIS equipment. Another major obstacle is the need for an MTC and centralised key management.

Kessler proposed a modification to the AIS protocol which is called Protected AIS (pAIS) [34]. The protocol offers protection against replay attacks (i.e., replaying legitimate AIS messages from the past) and allows to authenticate the sender and verify the message integrity. In addition, pAIS aims to be compatible with existing AIS equipment. The pAIS protocol involves creating a traditional AIS message and appending the so-called protect string to it. To build a protect string, the sender of the AIS message has to calculate a checksum for the message and create a timestamp string which reflects the current time, encrypting the concatenation of these values with its private key. If the pAIS message is received by a pAIS-aware device, it can use the public key of the sender to decrypt the protect string and verify the message integrity and creation time with the obtained checksum and timestamp. According to the author, the devices which are not aware of the pAIS protocol should be able to drop the protect string and process only the traditional AIS data that precedes it. However, as pointed out in [54], some receivers might drop not only the protect string but the entire message, as the presence of the protect string violates the AIS protocol standard.

3.3 "Special Cases" for Secure AIS

Encrypted AIS (EAIS) is a standard developed by US Coast Guard [53]. EAIS relies on AIS message types 6, 8, 25, and 26 that have been allocated for application-specific messages in the AIS standard. EAIS transmits the message header with sender and destination information in unencrypted form, whereas the remaining message is encrypted with AES (symmetric encryption algorithm that utilises a shared secret key). According to the EAIS standard, transceivers should also be able to accept and send regular unencrypted AIS messages if working in normal mode, while restricted mode involves sending EAIS messages only. Although EAIS is able to address many security issues of the AIS protocol, it has been designed for use by US governmental agencies.

Some commercial AIS products that use symmetric cryptography exist. Such a product is SAAB's R6 Secure AIS [44]. Nevertheless, they provide proprietary encrypted AIS only in vessels equipped with the same AIS product (e.g., Coast Guard and Police).

However, a wide adoption of a symmetric cryptography system that distributes the same symmetric keys on a vast number of vessels in a broad maritime context, as is the case for both solutions described above, is deemed to be a very complex task. Even more challenging is to keep the symmetric key secret after its distribution to a very large number of vessels [49].

Oh et al. [41] proposed an authentication scheme for vessels which involves the use of a TTP. The scheme assumes that each vessel pre-authenticates itself with the TTP before conducting mutual authentication with another vessel or group authentication with several vessels. The authors have not provided a detailed discussion of how the proposed scheme is further utilised for securing AIS traffic. Also, the scheme assumes that TTP is always available over the Internet which is not always the case for vessels at sea.

A study by Chen and Wu [15] proposed a vessel-based deep learning approach to detect sunken reefs and avoid collisions with them, transmitting information about sunken reefs to nearby ships with digitally signed AIS messages. The authors built a relevant prototype system and found it to work efficiently. However, the study does not focus on key management issues. Also, despite a successful prototype, adopting the proposed approach in the industry would require modification or replacement of existing AIS equipment.

4 Discussion

From the review of all papers in this chapter, four main obstacles for the adoption of the proposed solutions have been identified. For the case of backward-compatible solutions, the main obstacles are: (i) the induced overhead for broadcasting the signed messages leading to the overloading of the AIS-based communications, especially in high-traffic areas, and, (ii) the complexity of the solutions if alternative channels are to be used for the security-related messages, requiring additional system hardware and software components. For the case of non-backward-compatible solutions, the main obstacle is the requirement to update the AIS protocol standard by altering existing or adding new AIS message types. An additional requirement is the need to modify the AIS equipment or software. Such requirements are hard to implement in the maritime domain, especially for a broader application context. The fourth and relatively common obstacle for most of the proposed solutions, backward compatible and non-backward compatible, is the assumption that a PKI is organised and there is a TTP to manage the keys distribution. Again, this is something very difficult to implement in practice. Thus, adding security to AIS broadcasts has been shown to be problematic and rather infeasible, at least currently and in the near future.

A more feasible way forward is to put additional development effort only on lightweight backward-compatible solutions (assuming the PKI issue can be resolved), such as the ones proposed by Struck and Stoppe [50] and by Shyshkin [47]. Even if complete security cannot be guaranteed by future solutions, combining such approaches with real-time AIS cyberattack detection systems can provide the needed protection.

Recently, several papers have been published on analysing AIS messages to detect AIS spoofing and other attacks. For example, a study by Kontopoulos et al. [36] described a distributed architecture which was able to process large volumes of AIS messages in real time and where a master node divided the workload between many worker nodes. In order to detect spoofed AIS messages, the architecture tracked changes in vessel positions from consecutive AIS messages, reacting to anomalies (e.g., abnormally large position change in a short amount of time). The proposed architecture was evaluated on a large dataset of 43 million AIS messages.

Another example is the work by Iphar et al. [29] proposing a rule-based system for AIS anomaly detection which consisted of more than 900 rules. The rules were written by human experts and were divided into four classes by the AIS data they analysed: (1) one field from one AIS message was analysed, (2) several fields from one AIS message were analysed, (3) fields from several AIS messages of the same type were analysed, (4) fields from several AIS messages of all types were analysed. The rule-based system was used for assessing vessel-related risks, and the authors evaluated the system prototype on 24 million real-life AIS messages from 6-month time frame.

Louart et al. [39] developed a method for detection of AIS messages falsifications and spoofing by checking messages compliance with the TDMA communication protocol. The authors applied a Kalman filter to track every vessel and to assess the consistency of their velocity data sent because the vessel velocity can affect the TDMA protocol. The proposed method was validated on real data and showed promising results, being at the same time computationally cheap for real-time application.

Regarding the application of state-of-the-art machine learning (ML) techniques, Campbell et al. [13] compared several different ML techniques to identify the most suitable ones for the detection of AIS spoofing, motivated by a real event in the North Atlantic in April 2020 where more than 200 fake vessels appeared suddenly. The AIS data fields considered were the MMSI, date, time, speed over ground, course over ground, latitude, and longitude. The ML techniques investigated were K-means clustering, Decision Tree (DT), Random Forest (RF), Feed-Forward Neural Networks (FNN), Support Vector Machines (SVM), and One-Class Support Vector Machines (One-SVM). The results showed that DT, RF, and FNN best identified the fabricated AIS messages with F1 scores greater than 93 per cent on the test data.

It can be seen from the literature that the research topic of AIS data anomaly and spoofing detection has been gaining increasingly more attention, and the example works above show that it can be handled from various angles with good results. Hence, it is a natural way forward to integrate such solutions with security-enhancing AIS broadcast protocol approaches.

5 Conclusion

In this chapter, we reviewed all the published proposed solutions for providing secure AIS message broadcasts. We first categorised the approaches into backward compatible and non-backward compatible to the AIS protocol standard and then described the characteristics of each approach including insights on the implementation feasibility and possible shortcomings. Following the review of all the approaches, we provided an overall discussion on the main obstacles and assumptions for the implementation of such approaches, as well as we proposed a possible way forward.

Acknowledgment Research for this publication was funded by the EU Horizon 2020 project 952360-MariCybERA.

References

1. M. Afenyo, L.D. Caesar, Maritime cybersecurity threats: gaps and directions for future research. Ocean Coastal Manag. **236**, 106493 (2023). https://doi.org/10.1016/j.ocecoaman.2023.106493. https://www.sciencedirect.com/science/article/pii/S0964569123000182
2. AIS for Safety and Tracking: A Brief History - Global Fishing Watch. https://globalfishingwatch.org/article/ais-brief-history/. Accessed on 20 Sept 2024
3. A. Androjna, T. Brcko, I. Pavic, H. Greidanus, Assessing cyber challenges of maritime navigation. J. Marine Sci. Eng. **8**(10), 776 (2020). https://doi.org/10.3390/jmse8100776. https://www.mdpi.com/2077-1312/8/10/776
4. A. Aziz, P. Tedeschi, S. Sciancalepore, R.D. Pietro, SecureAIS- securing pairwise vessels communications, in *Proceedings of 2020 IEEE Conference on Communications and Network Security (CNS)* (IEEE, Piscataway, 2020), pp. 1–9. https://doi.org/10.1109/CNS48642.2020.9162320

5. M. Balduzzi, A. Pasta, K. Wilhoit, A security evaluation of AIS automated identification system, in *Proceedings of the 30th Annual Computer Security Applications Conference, ACSAC '14* (Association for Computing Machinery, New York, 2014), pp. 436–445. https://doi.org/10.1145/2664243.2664257
6. Baltic and International Maritime Council: The Guidelines on Cyber Security Onboard Ships. https://www.bimco.org/about-us-and-our-members/publications/the-guidelines-on-cyber-security-onboard-ships (2024). Accessed on 20 Sept 2024
7. E. Barker, Recommendation for key management. Special Publication 800-57, Part 1 Revision 5. National Institute of Standards and Technology, Gaithersburg, MD (2020)
8. M.A. Belokas, Maersk Line: Surviving from a cyber attack. [Online]. Available from: https://safety4sea.com/cm-maersk-line-surviving-from-a-cyber-attack/ (2018)
9. V. Bolbot, K. Kulkarni, P. Brunou, O.V. Banda, M. Musharraf, Developments and research directions in maritime cybersecurity: a systematic literature review and bibliometric analysis. Int. J. Critical Infrastruct. Protect. **39**, 100571 (2022). https://doi.org/10.1016/j.ijcip.2022.100571. https://www.sciencedirect.com/science/article/pii/S1874548222000555
10. C. Bronk, P. de Witte, *Maritime Cybersecurity: Meeting Threats to Globalization's Great Conveyor.* (Springer International Publishing, Cham, 2022), pp. 241–254. https://doi.org/10.1007/978-3-030-91293-2_10
11. C. Bueger, T. Liebetrau, Critical maritime infrastructure protection: what's the trouble? Marine Policy **155**, 105772 (2023). https://doi.org/10.1016/j.marpol.2023.105772. https://www.sciencedirect.com/science/article/pii/S0308597X23003056
12. F. Cabrera, N. Molina, M. Tichavska, V. Araña, Automatic identification system modular receiver for academic purposes. Radio Sci. **51**(7), 1038–1047 (2016). https://doi.org/10.1002/2015RS005895
13. J.N. Campbell, A.W. Isenor, M.D. Ferreira, Detection of invalid AIS messages using machine learning techniques. Procedia Comput. Sci. **205**, 229–238 (2022). https://doi.org/10.1016/j.procs.2022.09.024. https://www.sciencedirect.com/science/article/pii/S1877050922008894. 2022 International Conference on Military Communication and Information Systems (ICMCIS)
14. Y. Chen, Satellite-based AIS and its Comparison with LRIT. TransNav Int J Marine Navigat Safety Sea Transport **8**(2), 183–187 (2014). https://doi.org/10.12716/1001.08.02.02
15. M.Y. Chen, H.T. Wu, An automatic-identification-system-based vessel security system. IEEE Trans. Ind. Inf. **19**(1), 870–879 (2023). https://doi.org/10.1109/TII.2021.3139348
16. T. Eriksen, G. Høye, B. Narheim, B. Jensløkken Meland, Maritime traffic monitoring using a space-based AIS receiver. Acta Astronautica **58**(10), 537–549 (2006). https://doi.org/10.1016/j.actaastro.2005.12.016. https://www.sciencedirect.com/science/article/pii/S0094576506000233
17. A. Goudosis, S. Katsikas, Secure AIS with identity-based authentication and encryption. TransNav Int. J. Marine Navigat. Safety Sea Transport. **14**(2), 287–298 (2020). https://doi.org/10.12716/1001.14.02.03
18. A. Goudosis, S. Katsikas, Secure automatic identification system (SecAIS): proof-of-concept implementation. J. Marine Sci. Eng. **10**(6), 805 (2022). https://doi.org/10.3390/jmse10060805
19. A. Goudosis, T. Kostis, N. Nikitakos, Automatic identification system stated requirements for naval transponder security assurance, in *Proceedings of 2nd International Conference on Applications of Mathematics and Informatics in Military Sciences (AMIMS)* (2012)
20. A. Goudossis, S.K. Katsikas, Towards a secure automatic identification system (AIS). J. Marine Sci. Technol. **24**(2), 410–423 (2019). https://doi.org/10.1007/s00773-018-0561-3
21. Hackers Took 'Full Control' of a Container Ship's Navigation Systems for 10 Hours. https://rntfnd.org/2017/11/25/hackers-took-full-control-of-container-ships-navigation-systems-for-10-hours-ihs-fairplay/ (2017). Accessed on 12 Feb 2023
22. J. Hall, J. Lee, J. Benin, C. Armstrong, H. Owen, IEEE 1609 influenced automatic identification system (AIS), in *Proceedings of 2015 IEEE 81st Vehicular Technology Conference (VTC Spring)* (IEEE, Piscataway, 2015), pp. 1–5. https://doi.org/10.1109/VTCSpring.2015.7145867

23. IALA Guideline - An Overview of AIS. https://www.navcen.uscg.gov/sites/default/files/pdf/IALA_Guideline_1082_An_Overview_of_AIS.pdf. Accessed on 20 Sept 2024
24. IALA: Guideline G1117: VHF Dat Exchange System (VDES) Overview. 3rd edn. https://raw.githubusercontent.com/IALAPublications/Guidelines/main/G1117%20Ed3.0%20VHF%20Data%20Exchange%20System%20(VDES)%20Overview.pdf (2024). Accessed on 27 Sept 2024
25. International Maritime Organization: Resolution A.917(22) - Guidelines for the Onboard Operational Use of Shipborne Automatic Identification Systems (AIS) (2001)
26. International Maritime Organization: MSC-FAL.1-Circ.3 - Guidelines on Maritime Cyber Risk management (2017)
27. International Maritime Organization: SOLAS: Consolidated Text of the International Convention for the Safety of Life at Sea, 1974, and Its Protocol of 1988, Articles, Annexes and Certificates, Incorporating All Amendments in Effect from 1 January 2020 (2020). https://books.google.hu/books?id=JKULzgEACAAJ
28. International Maritime Organization: MSC-FAL.1/Circ.3/Rev.2 - Guidelines on Maritime Cyber Risk management (2022)
29. C. Iphar, A. Napoli, C. Ray, An expert-based method for the risk assessment of anomalous maritime transportation data. Appl. Ocean Res. **104**, 102337 (2020). https://doi.org/10.1016/j.apor.2020.102337. https://www.sciencedirect.com/science/article/pii/S0141118720304314
30. ITU-R: Technical characteristics for an automatic identification system using time division multiple access in the VHF maritime mobile frequency band. https://www.itu.int/dms_pubrec/itu-r/rec/m/R-REC-M.1371-5-201402-I!!PDF-E.pdf (2014). Accessed on 27 Sept 2024
31. M.H. Jahanian, F. Amin, A.H. Jahangir, Analysis of TESLA protocol in vehicular ad hoc networks using timed colored Petri nets, in *2015 6th International Conference on Information and Communication Systems (ICICS)* (IEEE, Piscataway, 2015), pp. 222–227. https://doi.org/10.1109/IACS.2015.7103231
32. S. Jegadeesan, M.S. Obaidat, P. Vijayakumar, M. Azees, SEAT: secure and energy efficient anonymous authentication with trajectory privacy-preserving scheme for marine traffic management. IEEE Trans. Green Commun. Netw. **6**(2), 815–824 (2022). https://doi.org/10.1109/TGCN.2021.3126618
33. E.P. Kechagias, G. Chatzistelios, G.A. Papadopoulos, P. Apostolou, Digital transformation of the maritime industry: a cybersecurity systemic approach. Int. J. Critical Infrastruct. Protect. **37**, 100526 (2022). https://doi.org/10.1016/j.ijcip.2022.100526. https://www.sciencedirect.com/science/article/pii/S1874548222000166
34. G.C. Kessler, Protected AIS: a demonstration of capability scheme to provide authentication and message integrity. TransNav Int. J. Marine Navigat. Safety Sea Transport. **14**(2), 279–286 (2020). https://doi.org/10.12716/1001.14.02.02
35. S. Khandker, H. Turtiainen, A. Costin, T. Hämäläinen, Cybersecurity attacks on software logic and error handling within AIS implementations: a systematic testing of resilience. IEEE Access **10**, 29493–29505 (2022). https://doi.org/10.1109/ACCESS.2022.3158943
36. I. Kontopoulos, G. Spiliopoulos, D. Zissis, K. Chatzikokolakis, A. Artikis, Countering real-time stream poisoning: an architecture for detecting vessel spoofing in streams of AIS data, in *2018 IEEE 16th Intl Conf on Dependable, Autonomic and Secure Computing, 16th Intl Conf on Pervasive Intelligence and Computing, 4th Intl Conf on Big Data Intelligence and Computing and Cyber Science and Technology Congress (DASC/PiCom/DataCom/CyberSciTech)* (2018), pp. 981–986. https://doi.org/10.1109/DASC/PiCom/DataCom/CyberSciTec.2018.00139
37. S. Levy, E. Gudes, D. Hendler, A survey of security challenges in automatic identification system (AIS) protocol, in *Cyber Security, Cryptology, and Machine Learning*. ed. by S. Dolev, E. Gudes, P. Paillier (Springer Nature Switzerland, Cham, 2023), pp.411–423
38. R.E. Litts, D.C. Popescu, O. Popescu, Authentication protocol for enhanced security of the automatic identification system, in *Proceedings of 2021 IEEE International Black Sea Conference on Communications and Networking (BlackSeaCom)* (IEEE, Piscataway, 2021), pp. 1–6. https://doi.org/10.1109/BlackSeaCom52164.2021.9527840

39. M. Louart, J.J. Szkolnik, A.O. Boudraa, Le Lann, J.C., F. Le Roy, Detection of AIS messages falsifications and spoofing by checking messages compliance with TDMA protocol. Digital Signal Process. **136**, 103983 (2023). https://doi.org/10.1016/j.dsp.2023.103983. https://www.sciencedirect.com/science/article/pii/S1051200423000787
40. A. Mnar, The story you aren't being told about Iran capturing two American vessels. https://www.mintpressnews.com/the-story-you-arent-being-told-about-iran-capturing-two-american-vessels/212937/ (2016). Accessed on 2 Dec 2023
41. S. Oh, D. Seo, B. Lee, S3 (Secure Ship-to-Ship) information sharing scheme using ship authentication in the e-navigation. Int. J. Security Appl. **9**(2), 97–110 (2015). https://doi.org/10.14257/ijsia.2015.9.2.10
42. L. Pilosu, A. Autolitano, D. Brevi, R. Scopigno, Exploring TV white spaces for the mitigation of AIS weaknesses, in *2015 IEEE Symposium on Communications and Vehicular Technology in the Benelux (SCVT)* (IEEE, Piscataway, 2015), pp. 1–6. https://doi.org/10.1109/SCVT.2015.7374243
43. L. Pilosu, A. Autolitano, D. Brevi, R. Scopigno, Prototyping of an integrated, white-space solution for the enforcement of ship reporting systems, in *2015 Advances in Wireless and Optical Communications (RTUWO)* (IEEE, Piscataway, 2015), pp. 146–149. https://doi.org/10.1109/RTUWO.2015.7365739
44. SAAB: R6 Secure AIS. https://www.saab.com/products/r6-secure-ais (2024). Accessed on 27 Sept 2024
45. A. Saleem, H. Turtiainen, A. Costin, T. Hämäläinen, Backward-compatible software upgrades for ADS-B and AIS to support ECDSA-secured protocols, in *Proceedings of the 23rd European Conference on Cyber Warfare and Security, ECCWS 2024*. Academic Conferences International Limited (2024). https://doi.org/10.34190/eccws.23.1.2250
46. S. Sciancalepore, P. Tedeschi, A. Aziz, R. Di Pietro, Auth-AIS: secure, flexible, and backward-compatible authentication of vessels AIS broadcasts. IEEE Trans. Dependable Secure Comput. **19**(4), 2709–2726 (2022). https://doi.org/10.1109/TDSC.2021.3069428
47. O. Shyshkin, Cybersecurity providing for maritime automatic identification system, in *2022 IEEE 41st International Conference on Electronics and Nanotechnology (ELNANO)* (IEEE, Piscataway, 2022), pp. 736–740. https://doi.org/10.1109/ELNANO54667.2022.9926987
48. A. Stewart, E. Rice, P. Safonov, Digital authentication strategies for the automated identification system, in *Proceedings of the Midwest Instruction and Computing Symposium. MICS* (2018)
49. M. Strohmeier, V. Lenders, I. Martinovic, On the security of the automatic dependent surveillance-broadcast protocol. IEEE Commun. Surveys Tutor. **17**(2), 1066–1087 (2015). https://doi.org/10.1109/COMST.2014.2365951
50. M.C. Struck, J. Stoppe, A backwards compatible approach to authenticate automatic identification system messages, in *Proceedings of 2021 IEEE International Conference on Cyber Security and Resilience (CSR)* (IEEE, Piscataway, 2021), pp. 524–529. https://doi.org/10.1109/CSR51186.2021.9527954
51. P. Su, N. Sun, L. Zhu, Y. Li, R. Bi, M. Li, Z. Zhang, A privacy-preserving and vessel authentication scheme using automatic identification system, in *Proceedings of the Fifth ACM International Workshop on Security in Cloud Computing, SCC '17* (Association for Computing Machinery, New York, 2017), pp. 83–90. https://doi.org/10.1145/3055259.3055261
52. B. Svilicic, J. Kamahara, M. Rooks, Y. Yano, Maritime cyber risk management: an experimental ship assessment. J. Navigat. **72**(5), 1108–1120 (2019). https://doi.org/10.1017/s0373463318001157
53. USCG: Encrypted Automatic Identification System (EAIS) Interface Design Description (IDD). Command, Control, and Communications Engineering Center (C3Cen). https://epic.org/foia/dhs/uscg/nais/EPIC-15-05-29-USCG-FOIA-20151030-Production-2.pdf (2014). Accessed on 27 Sept 2024

54. G. Wimpenny, J. Safar, A. Grant, M. Bransby, Securing the automatic identification system (AIS): using public key cryptography to prevent spoofing whilst retaining backwards compatibility. J. Navigat. **75**(2), 333–345 (2022). https://doi.org/10.1017/S0373463321000837
55. G. Wimpenny, J. Safar, A. Grant, M. Bransby, N. Ward, Public key authentication for AIS and the VHF data exchange system (VDES), in *Proceedings of the 31st International Technical Meeting of the Satellite Division of the Institute of Navigation (ION GNSS+ 2018)* (ION Instutute of Navigation, Virginia, 2018), pp. 1841–1851. https://doi.org/10.1145/1321631.1321685

Open Access This chapter is licensed under the terms of the Creative Commons Attribution-NonCommercial-NoDerivatives 4.0 International License (http://creativecommons.org/licenses/by-nc-nd/4.0/), which permits any noncommercial use, sharing, distribution and reproduction in any medium or format, as long as you give appropriate credit to the original author(s) and the source, provide a link to the Creative Commons license and indicate if you modified the licensed material. You do not have permission under this license to share adapted material derived from this chapter or parts of it.

The images or other third party material in this chapter are included in the chapter's Creative Commons license, unless indicated otherwise in a credit line to the material. If material is not included in the chapter's Creative Commons license and your intended use is not permitted by statutory regulation or exceeds the permitted use, you will need to obtain permission directly from the copyright holder.

Using Incremental Inductive Logic Programming for Learning Spoofing Attacks on Maritime Automatic Identification System Data

Aboubaker Seddiq Benterki, Gabor Visky, Jüri Vain, and Leonidas Tsiopoulos

1 Introduction

Maritime transport has significant role in the global economy [1]; however, it is vulnerable to disruptions impacting severely whole global chains of trade [2]. Therefore, academia, industry, and public sector have recognised the need to improve the sector's resilience equalising its importance with other critical aspects in the maritime domain [3–7].

In 2017 and 2022, the IMO released guidelines with high-level recommendations on maritime cyber-risk management to promote effective cyber-risk management and protect the shipping industry from cyber-threats and vulnerabilities [8, 9].

International Ship and Port Facility Security (ISPS) Code established by the IMO partly regulates the cybersecurity risk assessment [10]. These regulations do not specify particular characteristics of cybersecurity solutions on a ship.

In June 2017, the Maritime Safety Committee also adopted the resolution for Maritime Cyber-Risk Management in Safety Management Systems to encourage administrations to appropriately address cyber-risks in existing safety management systems [11].

One of the key components to assure situation awareness at sea is the Automatic Identification System (AIS) [12, 13]. Vessels outfitted with AIS continuously transmit their unique identification along with other vital navigation data. Despite the criticality of AIS, its communication protocol is not secured from cyberattacks. This, in turn, makes AIS a desirable target of cyber-crime by altering or falsifying data and allowing the spread of inaccurate information about vessels [14–17].

A. S. Benterki (✉) · G. Visky · J. Vain · L. Tsiopoulos
Department of Software Science, Tallinn University of Technology, Tallinn, Estonia
e-mail: aboubaker.benterki@taltech.ee; gabor.visky@taltech.ee; juri.vain@taltech.ee; leonidas.tsiopoulos@taltech.ee

© The Author(s) 2025
S. Bauk (ed.), *Maritime Cybersecurity*, Signals and Communication Technology, https://doi.org/10.1007/978-3-031-87290-7_7

Timely detection of anomalies in the transmitted data can help the identification of such attacks and trigger adequate countermeasures.

Current research literature increases rapidly in the number of approaches focusing on anomaly detection in AIS data, with approaches ranging from statistical and rule-based [18–22], to exploiting neural network (NN)-based machine learning (ML) methods [23, 24]. However, there is an alternative ML methods group, based on inductive logic programming (ILP), introduced by Muggleton [25], that surprisingly has not got any attention in existing works, though ILP has demonstrated a series of advantages over other ML methods. The advantages include the ability to generalise from small numbers of training examples, natural support to lifelong and transfer learning, ability to learn complex relational theories, and explainability of learning results [26].

To our best knowledge, in this chapter, ILP is applied for the first time for anomaly detection in real-life AIS data. According to our experimental results on real AIS data collected at the harbour of Tallinn, we demonstrate that ILP extended with the support for learning also numerical relations can efficiently learn various AIS attack signatures and generalise them in the form of a compact set of attack detection rules.

The rest of this chapter is organised as follows. In Sect. 2 we present preliminaries of AIS, ILP, and an extension of ILP that we use in this chapter to learn logic programs combining relational logic and numerical reasoning. In Sect. 3 we discuss the related works. In Sect. 4 we present our method, and in Sect. 5 we evaluate it on a real-life dataset. In Sect. 6 we conclude this chapter and discuss the future work.

2 Preliminaries

2.1 Automatic Identification System (AIS)

The AIS is a coastal tracking system designed for short-range monitoring, typically effective up to a distance of 20–100 nautical miles (NM) at sea, depending on the setup [27, 28]. It was created to offer identification and positioning data to vessels and coastal stations, enabling them to monitor, identify, and share information about marine traffic. The International Maritime Organization (IMO), under the Safety of Life at Sea (SOLAS) Convention, mandates that AIS transponders be installed on international voyaging ships with a gross tonnage (GT) of 300 or more, all ships with a GT of 500 or above, and on all passenger ships irrespective of size. This requirement aims to enhance safety and navigation efficiency at sea [29].

AIS transponders are available in two different classes. Class A transponders are required for all SOLAS-compliant vessels, as previously mentioned. Class B transponders, on the other hand, are designed for non-SOLAS vessels, including domestic commercial vessels and pleasure crafts.

Table 1 Default timing of AIS messages [30]

Type of ship	Reporting interval
Ship at anchor	180 sec
Speed 0–14 knots	12 sec
Speed 0–14 knots and changing course	4 sec
Speed 14–23 knots	6 sec
Speed 14–23 knots and changing course	2 sec
Speed >23 knots	3 sec
Speed >23 knots and changing course	2 sec

The installed AIS transponder regularly broadcasts information on the ship's status, such as static details, dynamic (e.g., vessel position, speed, navigational status) data, and voyage (destination port and the estimated time of arrival of the vessel) information [30]. The dynamic AIS data is automatically transmitted every 2–10 seconds, depending on the vessel's speed, as detailed in Table 1. The static data is transmitted every 6 minutes regardless of the vessel's movement speed or status [31].

2.2 Inductive Logic Programming

ILP studies learning from examples, within the framework provided by clausal logic. The examples and background knowledge are given as clauses, and the theory that is to be induced from these also consists of clauses. ILP uses logic programming as a uniform clausal representation for examples, background knowledge, and hypotheses learned. Given an encoding of the known background knowledge and a set of examples represented as a logical database of facts, an ILP system derives a hypothesised logic program which entails all the positive and none of the negative examples. Formally:

- Given a finite set of clauses B (background knowledge), and sets of clauses $E+$ and $E-$ (positive and negative examples, respectively).
- Find a theory Σ, such that $\Sigma \cup B$ is correct with respect to $E+$ and $E-$.

By the correctness of theory Σ we mean that $\Sigma \cup B \models e+, \forall e+ \in E+$ (*completeness*), and $\Sigma \cup B \not\models e-, \forall e- \in E-$ (*consistency*). The two basic steps in the search for a correct theory are *specialisation* and *generalisation*. If the current theory together with the background knowledge entails some of the negative examples, it is too strong and needs weakening, i.e., specialisation, such that the new theory and the background knowledge are consistent with respect to the negative examples. If the current theory together with the background knowledge does not imply all positive examples, it needs to be strengthened (generalised) by finding a more general theory such that all positive examples are implied.

In ILP setting we now formulate the learning task as the task of learning rules that generalise correct AIS data exchange, and thus by monitoring AIS data and checking them against learned rules (in terms of ILP theory), it allows distinguishing anomalies from normal AIS data. This, in turn, could indicate the possibility of cyber-incidence, e.g., spoofing attacks. As a case study, the positive examples are based on a small (but sufficient) subset of real AIS data collected at the harbour of Tallinn, and the negative examples are based on extracted samples from the same subset of data which violate the accuracy constraint for the reported consecutive vessel positions. Additionally, we injected fake vessels into this dataset to simulate spoofing attacks (see Sect. 4.2.3).

2.3 Relational Program Synthesis with Numerical Reasoning (NUMSYNTH)

NUMSYNTH [32] is an ILP framework designed to learn programs that combine relational logic and numerical reasoning. While many ILP frameworks primarily focus on learning Prolog programs with symbolic reasoning, NUMSYNTH extends this by learning programs with numerical values, crucial for tasks involving continuous domains like real numbers or discrete domains of integers. NUMSYNTH ensures that for each positive example, a learned hypothesis includes numerical constraints that cover the positive examples. In contrast, for each negative example, the hypothesis excludes the values that would entail it. NUMSYNTH incorporates this form of reasoning by using two stages: program search and numerical search. In the first stage, partial programs are generated with numerical variables. In the second stage, satisfiability modulo theories (SMT) solvers are employed to search for appropriate numerical values that fit the training data. This two-stage process allows NUMSYNTH to handle infinite numerical domains and derive numerical thresholds, constraints, or inequalities from multiple examples. The system's ability to reason over examples jointly, rather than individually, distinguishes it from other ILP systems that struggle with complex numerical relationships.

3 Related Work

Since the AIS system is not designed to be resistant against cyberattacks, it brings severe vulnerability into the Vessel Traffic Service systems. Balduzzi et al. [33, 34] introduced the security consideration related to this system. Due to this, a lot of research has been conducted to identify anomalies in AIS data, including anomalies possibly caused by cyberattacks.

Several research works focused on the identification of anomalies in ship trajectories by applying different methods. Ristic et al. [35] used statistical analysis of the position of the ships for anomaly detection in trajectory, with promising results; however, their method generated high false positive alerts.

Kowalska and Peel [36] used data-driven, non-parametric Bayesian model with active learning for anomaly detection in ships' trajectory. Vespe et al. [37] applied unsupervised learning for anomaly detection in maritime traffic patterns, using real-time and historical AIS data. The method successfully identified ships violating the traffic separating zones and prohibited areas.

Katsilieris et al. [38] studied the trustworthiness of AIS data with the help of radar measurements and information from the tracking system. The applied log-likelihood ratio test delivered good results, especially if the data from the AIS and RADAR system deviated enough.

Coleman in his thesis applied and analysed several different ML-based anomaly detection methods on the heatmap of ships' location based on data [39].

Kontopoulos et al. [40] studied the detection of data spoofing and falsification attacks in real-time environment. Their method determines the average speed needed to travel along the shortest route between two consecutive locations reported via AIS. If the computed speed falls within a realistic range, the next message is accepted as the new last valid position for that vessel. If not, the message is marked as potentially spoofed. With this approach, they achieved moderate results.

Kullberg et al. [41] developed a method that recursively learned a model of the nominal vessel routes from AIS data and simultaneously estimated the current state of the vessels. The method also distinguished anomalies and measurement outliers. Statistical testing relative to a current motion model was applied, and the method was evaluated against historical AIS data showing that previously unseen motions could be detected.

d'Afflisio et al. [42] proposed an anomaly detection strategy based on a multiple hypothesis testing framework using two approaches. The first approach was based on the generalised likelihood ratio testing, and the second approach was based on the model-order selection methodology applying an appropriate penalty term to the maximised log likelihood based on the statistical model for the vessel kinematic. Anomaly detection rules were then derived, and the effectiveness of the approach was demonstrated against simulated data.

All the approaches discussed above brought good results; however, they did not consider any further characteristics of the AIS transmissions, like transmission periodicity, etc.

Lane et al. [43], besides the trajectory-related anomalies, considered the unexpected AIS activity as anomalous behaviour focusing on the *existence* of the transmissions. The approach generated false positives, if a signal was received from a not covered area, so the authors improved this method by building a receiver coverage map. This approach works with land-based receivers, but on the sea its usefulness is questionable. To determine the probability of a higher-level threat, a general Bayesian network-based method was used.

Iphar et al. [22] proposed a rule-based method for AIS data integrity assessment, with rules derived manually from the system's technical specifications and with help by domain experts. The study focused on the different characteristics of the transmitted data, like consistency, next position violation, (dis)appearing of a transmitter, etc. 666 of these rules were implemented in Python as part of an AIS data anomaly detection component.

Blauwkamp et al. analysed 334 million AIS messages. They conducted statistical analysis, supervised classification, and unsupervised clustering for feature selection and proposed the implementation of a behaviour-based anomaly detection system based on leveraging deep neural networks, historical AIS data, and logic rules [44]. Their method uses message types, location, velocity, and other attributes with known behaviour for training the model to identify deviations in message patterns. Another indicator of aberrant activity is a deviated or unknown response sequence to a base station's interrogation.

Louart et al. [45] developed a method for detection of AIS messages falsifications and spoofing by checking messages compliance with the time-division multiple access (TDMA) communication protocol, which is the protocol employed for AIS data broadcasts. The authors applied a Kalman filter to track every vessel and to assess the consistency of their velocity data sent because the vessel velocity can affect the TDMA protocol. The proposed method was validated on real data and showed promising results, being at the same time computationally cheap for real-time application. Importantly, the authors provided open access to the source codes to foster research activities from both industry and academia in this field.

Louart et al. [46] also developed an approach that detects AIS identity spoofing combining the tracking of the vessel position and AIS transceiver's carrier frequency offset caused by the carrier frequencies mismatch between emitter and receiver and Doppler effect. This offset is used as a radiometric signature to identify every transceiver independently of its transmitted identity. The offset can drift over time and, thus, it is tracked by a Kalman filter. Vessel position is also considered to reduce the miss probability of spoofing detection. The method was tested on real AIS data, and the results demonstrated very low false alarm (1%) and miss probabilities (1.7%). The algorithm and AIS data used are open access.

Similarly to the work by Coleman [39] that compared different ML techniques for trajectory anomaly detection, Campbell et al. [23] compared several different ML techniques to identify the most suitable ones for the detection of AIS spoofing, motivated by a real event in the North Atlantic in April 2020 where more than 200 fake vessels appeared suddenly. The AIS data fields considered were the MMSI, date, time, SOG, COG, latitude, and longitude. Data cleaning involved removal of duplicate messages and of records that contained physically invalid entries. The final dataset covered April 1–30, 2020, and contained 19,029 data entries. The dataset had 17,853 entries that belonged to the valid vessel class and 1176 entries as part of the invalid vessel class. The data set was divided into training, validation, and test sets with a 60%/20%/20% split, respectively. The ML techniques investigated were K-means clustering, Decision Tree (DT), Random Forest (RF), Feed-Forward Neural Networks (FNN), Support Vector Machines (SVM), and One-Class Support Vector Machines (One-SVM). The results showed that DT, RF, and FNN best identified the fabricated AIS messages with F1 scores greater than 93 per cent on the test data.

For an additional comprehensive review on supervised and unsupervised ML techniques to detect abnormal activities, behaviours, and intents in AIS data, the reader is referred to the review paper by Gamage et al. [24].

Based on our review of existing literature, various methods have been employed for anomaly detection in identifying attacks against AIS, each yielding different strengths and weaknesses. Compared to the ML- and rule-based approaches described above, our method requires much less AIS data to automatically learn logically correct-by-construction anomaly detection rules, and it does not require a complex mathematical model of the vessel kinematics nor extensive expert involvement to guide the feature extraction and efficient rule formation. Furthermore, our method contributes to Explainable AI with human understandable "if-then" type rules.

4 Learning Method for AIS Attack Detection

General flowchart describing the main steps of applying ILP for learning rules of AIS attack detection is depicted in Fig. 1. In the following sections, each of the steps is described in detail.

4.1 Data Collection and Preprocessing

There are publicly available AIS data from Marrinetraffic [47], Spire Global [48], or AISHub [49]. These sources make valuable contributions but focus mainly on the ships' trajectory, position, and movement. As far as only aggregated data is available from these sources [50], these datasets are not fine-grained enough to support detailed AIS anomaly detection-related research. To assure the trustworthiness of our solution, we used for training and testing it on a real-life MarCyb dataset, collected with a dedicated receiver installed on the premises of Estonian Maritime Academy, N59.462N, E24.666 40m ASL [51, 52].

The AIS data used in this study was recorded in the Baltic Sea. For this experiment, we extracted a portion of the data from **19/06/2022** at **2:00:00** to **20/06/2022** at **2:00:00**, providing a 24-hour snapshot of vessels movements in the area.

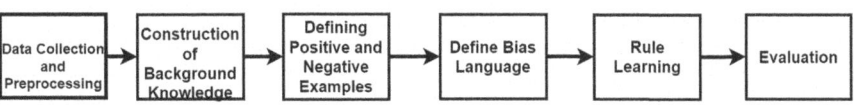

Fig. 1 Flowchart of learning AIS attack signatures using ILP

Vessels, passed the edge of the AIS receiver's coverage, appeared only a few times in the dataset. During the training set's preprocessing, we excluded these vessels' data if only one or two messages within a normal reporting interval (less than 3 minutes) were received from them. These vessels were removed as outliers causing unnecessary noise in the dataset since they prevent accurate behaviour analysis and rule derivation. The remaining vessels' data was used for the learning.

In the data cleaning phase, and in order to answer the reliability question, messages including data fields with default values (such as speed value of 102.3 knots) were excluded due to being uninformative. We also excluded messages with fields inserted by the crew and fields such as RAIM and accuracy, which were found to be inconsistent with AIS specifications. The cleaned dataset was the basis for constructing background knowledge and positive examples for the ILP system to learn general rules describing correct AIS data.

4.2 Learning Rules of Correct AIS Data

4.2.1 Setup of the NUMSYNTH Environment

In ILP, learning is structured around constructing three key components: *background knowledge*, *examples*, and *bias language*. NUMSYNTH adopts this approach to derive meaningful rules from data, and we followed the same methodology when learning rules describing normal vessel behaviour from AIS data.

4.2.2 Constructing Background Knowledge

In ILP, background knowledge plays a key role in guiding the system during the learning process. It consists of facts and known information that the ILP system uses to derive rules. For this experiment, the background knowledge is derived from the AIS message type 1 data, containing the ground truth for each recorded vessel. The key components of background knowledge include:

- *The number of messages* encoded as Prolog fact (msg_nbr/2) representing the total number of AIS messages reported by a vessel.
- *Duration* encoded in the fact (duration/2) that specifies the time interval (measured in seconds) when the vessel's AIS has been received.
- *Maximum and minimum speed* (max_speed/2, min_speed/2), the highest and lowest speeds reported by the vessel.
- *Mean and standard deviation of position error*, calculated based on two factors: According to AIS specifications, the coordinate accuracy is within 10 m if the accuracy field is set to true. To account for this, we calculate the difference between the distance derived from longitude and latitude between two coordinates reported in consecutive messages and the distance computed using the

reported time and speed. From now on, we call this distance difference *delta distance*. Since we are dealing with two points, the accuracy of calculated distance can be assumed to be less than 20 m. Given that each vessel has multiple reported messages, we calculate the ground truth of the positional error by determining the mean and standard deviation in two ways. The first method includes only the points where the delta distance, as described above, is 20 m or less. The second method includes all data points, regardless of whether the delta distance is greater or less than 20 m. We encode the mean and standard deviation for the points with a delta distance of 20 m or less as (mean/2) and (standard_deviation/2), and for the second case, we use (mean1/2) and (standard_deviation1/2).

Our analysis shows that whether the accuracy field is set to true or false, most of the data points still meet the accuracy threshold expected.

- Predicate (negative/2) is used to count the number of messages that exceeded the delta distance constraint.
- Predicate (status/2) indicates whether the vessel is moving or stopped based on its reported speed. A vessel is considered to be in a stopped state if its maximum speed was less than or equal to 0.4 knots.
- Predicate(trajectory/2) determines whether a vessel follows a straight or non-straight trajectory.

The background knowledge is expressed as a set of Prolog facts where fact describes a specific characteristic of the vessel. An example of background knowledge for a given vessel (sh_276636222) is as follows:

```
msg_nbr(sh_276636222, 66).
duration(sh_276636222, 1241.23).
max_speed(sh_276636222, 17.5).
min_speed(sh_276636222, 17.1).
mean(sh_276636222, -10000000.0).
standard_deviation(sh_276636222, -10000000.0).
negative(sh_276636222, 66).
mean1(sh_276636222, 148.77).
standard_deviation1(sh_276636222, 2.13).
status(sh_276636222, moving).
trajectory(sh_276636222, straight).
```

This background knowledge, combined with the examples and bias declarations, allows NUMSYNTH to explore the space of possible hypotheses and derive rules that explain normal vessel behaviour.

4.2.3 Defining Positive and Negative Examples

Defining positive and negative examples (sets E+ and E−, respectively) is a critical step for getting correct learning hypothesis. We defined **negative examples** as vessels that have sent data violating the accuracy constraint (i.e., a delta distance larger than 20 m between positions reported in consecutive messages). The violations were caused by one or more of the following issues:

- Reporting position outside the zone where the data was collected
- Significant speed jumps reflected in consecutive messages
- Long time gaps between consecutive message transmissions

Any vessel with one or more of these violations was labelled as a negative example. On the other hand, positive examples were the vessels data that did not violate any of these constraints. These vessels adhered to the expected behaviour, maintaining a consistent position and speed, and thus represent normal behaviour in the dataset. In addition to the naturally occurring negative examples, we additionally generated *fake vessels* to imitate spoofing attacks. These synthetic data about the vessels were designed with two types of trajectories: one with a straight trajectory and another with a circular trajectory, to simulate suspicious or abnormal behaviour. To make these synthetic examples more realistic, we:

- Deleted certain data points to simulate missing transmissions
- Extracted segments from real vessel data (such as speed) to mimic normal conditions, while modifying the time and trajectory information to introduce anomalies

4.2.4 Language Bias and Search Space

The NUMSYNTH *bias language* is a declarations file that informs the system which predicates from the background knowledge can be used during the learning process. This file defines the structure of the rules that NUMSYNTH can generate, specifying which predicates are allowed in the head and which ones in the body of the resulting rules, as well as the types and directions (input, output) of the variables involved. It also provides information to limit the search space, such as the type of variables and the direction (input or output) for each predicate used in rule generation, and how many numerical variables must be bounded for a valid rule. An example of the bias file used in this study is as follows:

```
head_pred(vessel_id,1).
body_pred(min_speed,2).
type(min_speed,(string,float)).
direction(min_speed,(in,out)).
```

- `head_pred/2`: Declares the head of the resulting rule, which, in this case, is the predicate `fact/1`. The number 1 indicates that the predicate has arity 1.
- `body_pred/2`: Declares a predicate that can appear in the body of the learned rule. For example, `min_speed/2` is a predicate with an arity 2, which can be used to describe the minimum speed of a vessel.
- `type/2`: Specifies the types of the variables involved in the predicate. For the `min_speed/2` predicate, the first variable is of type string (representing the

vessel identifier), and the second variable is of type float (representing the minimum speed).
- `direction/2`: Indicates the direction of the variables. `in` means the variable is an input (e.g., the vessel identifier), while `out` means the variable is an output (e.g., the minimum speed value calculated by the system).

This bias file plays a crucial role in guiding the ILP system by:

- Defining the *structure* of the rules that can be generated, ensuring that only meaningful rules are created based on the available data.
- Restricting the *search space* by limiting the predicates and their types, which improves computational efficiency.
- Indicating how variables flow through the rule: Input variables (in) are provided from the data, while output variables (out) are inferred by the system.

NUMSYNTH uses the so-called *generate, test, and constrain* approach [53], where it generates rules based on these declarations, tests them against the examples, and constrains the search space iteratively. The system begins with rules of size 1 and increases the size of the rule incrementally while respecting the variable constraints. The search process continues until a valid rule is found or the maximum size limit is reached, ensuring that the system efficiently explores possible hypotheses without getting stuck in an exhaustive search.

4.2.5 NUMSYNTH Derived Rules

Inductive Logic Programming (ILP) aims to discover minimal, non-disjunctive rules that accurately classify all positive examples while excluding all negative ones. Using NUMSYNTH, we applied a systematic refinement approach to optimise the generated rules over three learning epochs. Each epoch produced increasingly concise and accurate rules to classify vessel behaviour based on AIS data, culminating in perfect classification.

4.3 Summary of Results

The performance metrics across the three learning epochs are summarised in Table 2:

Table 2 Summary of learning epochs: precision, recall, and rule size

Epoch	Precision	Recall	TP	FN	TN	FP	Rule size
Epoch 1	1.00	0.61	27	17	34	0	16
Epoch 2	1.00	1.00	44	0	34	0	9
Epoch 3	1.00	1.00	44	0	34	0	6

Through these iterations, the rules were refined to achieve both precision and recall scores of 1.00, with the final phase producing the most concise and effective rule set. These rules provide a reliable classification of vessel behaviour, leveraging key features such as trajectory, status, and numerical values like mean value and standard deviation of speed.

4.4 Learning Epochs in Detail

4.4.1 Epoch 1

In the initial iteration, the system was configured with a maximum of seven variables and one numerical variable. Despite an early system crash due to inconsistencies between positive and negative examples, we deactivated conflicting predicates to resolve these issues. This led to the following rules being generated:

```
vessel_id(A) :- standard_deviation1(A, C),geq(C,
                0.254), standard_deviation(A, C).
vessel_id(A) :- standard_deviation(A, C),
                trajectory(A, straight), geq(C, 3.089).
vessel_id(A) :- status(A, stop), mean(A, B), geq(B,
                0.274).
vessel_id(A) :- standard_deviation(A, D), leq(D,
                0.439), status(A, stop).
```

Although *precision* reached 1.00, the recall was only 0.61, indicating that 17 positive examples were missed, and the rule size was relatively large (16 literals).

4.4.2 Epoch 2

In Epoch 2, we increased the number of numerical variables to two, enabling the system to explore more complex relationships. This adjustment improved recall to 1.00, achieving perfect classification with the following rules:

```
vessel_id(A) :-standard_deviation1(A,B),
               leq(B,17.564),standard_deviation(AC),
               geq(C, 0.222).
vessel_id(A) :- standard_deviation1(A, C),status(A,
                stop), standard_deviation(A, C).
```

The system successfully classified all examples with a smaller rule size (nine literals), demonstrating the advantage of incorporating additional numerical variables.

4.4.3 Epoch 3

In Epoch 3, we further increased the number of numerical variables to three, yielding a more concise and effective rule set. The rules produced in this phase were:

```
vessel_id(A) :- standard_deviation1(A, E),mean(A,
                D),geq(D, 0.012),geq(E, 0.084),
                leq(E,17.564).
```

This epoch resulted in both *precision* and *recall* scores of 1.00, while reducing the rule size to six literals. The system successfully generated a rule that can distinguish between normal and abnormal vessel behaviour based on the bounded values of the parameters in predicates mean and standard_deviation1.

4.4.4 Learning Runtime

The learning process of Epoch 3 (with three numerical variables) took the longest time, approximately **15** seconds, running on an Intel Core i5 processor. This demonstrates the feasibility of NUMSYNTH in handling complex numerical reasoning tasks, even with multiple variables and large datasets, and the scalability of our approach for real-time and data-intensive applications.

4.4.5 Saturation of the Learning Process

When the number of numerical variables was increased beyond 3, the solution remained the same as with three variables. This indicates that three numerical variables were sufficient to capture the complexity of the dataset required to distinguish between normal and abnormal vessel behaviour based on AIS data. Further increases did not contribute to further optimisation of the rule set.

5 Evaluation of the Method

In this section, we evaluate the method by studying the results of learned rules based on three criteria: (1) validity of the rules, (2) explainability, and (3) the limitations of using only AIS Message Type 1 to detect spoofing attacks.

5.1 *Validity of the Rules*

The rules learned by NUMSYNTH were tested using a snapshot of AIS data that are independent from the training dataset. The learned rules were valid in correctly classifying both normal and abnormal vessel behaviour. This validation was based on real-world AIS data collected over a 24-hour period from the Baltic Sea, specifically in the Tallinn Bay area. The rules demonstrated perfect classification performance within the dataset, achieving 100% in both precision and recall.

However, one has to admit that our current validation dataset is somewhat limited. Though Tallinn Bay is one of the most intensive traffic zones in Baltic Sea, the dataset does not capture the full range of global maritime conditions, such as varying traffic density, environmental conditions, or regional navigation patterns. Testing the learned rules with additional datasets from different regions would allow us to confirm the method feasibility in a broader context. By using data from other maritime regions, we can determine whether the rules are robust enough to handle different vessel behaviours and operational patterns across the world.

In addition to expanding the dataset, future work will focus on determining the perfect time interval for testing the rules. The perfect time interval refers to the interval that best captures the vessel's position during different scenarios, such as when it is moving or stopping. This is important because the behaviour of a vessel may vary significantly depending on its status. Furthermore, we must consider the coverage area, as vessels may still appear in the dataset after reaching their intended destination and starting a new trip, thus creating a new scenario that requires different analyses. Better selection of data collecting time intervals will help improve the accuracy and consistency of the rule extraction and application across various vessel operations.

5.2 Explainability

A significant advantage of using ILP is the explainability of the learned rules. The rules generated are interpretable and can be effortlessly understood by domain experts. For instance, rules based on vessel trajectory, speed, and positional accuracy are intuitively connected to maritime operations and can be reviewed and explained easily. This explainability makes the system more transparent and trustworthy, as users can inspect the specific conditions under which a vessel is flagged as behaving abnormally.

In comparison to black-box ML models, the ability to review and explain the rules ensures that the system can be used confidently in operational settings. It also allows for continuous improvement and adaptation, as new data and insights from maritime experts can be integrated during the rule refinement process, thus carrying out lifelong learning approach.

5.3 Limitations of Using Only AIS Message Type 1

While the rules based on AIS Message Type 1 data effectively distinguish between normal and abnormal vessel behaviour, they still are limited in their ability to definitively detect spoofing attacks. AIS Message Type 1 contains dynamic information such as vessel position, speed, and course, which is helpful for identifying deviations from expected behaviour. However, without additional data sources (e.g., AIS Message Types 5 or 24, radar data), it is impossible to determine with full certainty whether abnormal behaviour is caused by a spoofing attack or by other factors, such as technical malfunctions or human error.

In summary, the current system can detect when a vessel is behaving abnormally, but it cannot guarantee that the abnormality is due to a cyberattack like spoofing. To accurately identify spoofing, the system would need to incorporate other AIS message types or external verification systems (e.g., radar or satellite data). However, current approach still helps narrowing down the number of suspicious cases that require further analysis. Developing a more comprehensive detection system is part of our future work.

6 Conclusion and Future Work

This study demonstrates promising results of applying inductive logic programming method for learning anomalies in AIS data that could indicate potential spoofing attacks. The method has been successfully applied to real-life data monitored from the Baltic Sea area and used to learn human interpretable rules that distinguish normal and abnormal vessel behaviours. The key advantage of this approach is its ability to generalise from a small set of examples and provide explainable evidence of anomaly occurrence, making it suitable for operational use in maritime environments.

The learned rules achieved perfect precision and recall in classifying vessel behaviour to correct and anomalous based on real AIS data. However, the dataset used for validation of the learned rules is limited in this study to the Tallinn Bay area, which presents a somewhat limited range of global maritime conditions.

Future work will involve testing the relevance of the demonstrated method on a broader range of datasets and environments to ensure its robustness and range of applicability. As concluded from this study, improving the detection capability, specifically for reliable detection of spoofing attacks, may require extension of the training set with other AIS message types and with radar or satellite data. Nevertheless, this study opens new possibilities for future work in applying ILP methods for maritime cybersecurity, offering a step toward a more resilient and secure AIS system.

Acknowledgment Research for this publication was funded by the EU Horizon 2020 project 952360-MariCybERA.

References

1. H.N. Psaraftis, *The Future of Maritime Transport* (Elsevier, Amsterdam, 2021), pp. 535–539. [Online]. Available: https://doi.org/10.1016/b978-0-08-102671-7.10479-8
2. M.A. Belokas, Maersk Line: Surviving from a cyber attack, [Online]. Available from: https://safety4sea.com/cm-maersk-line-surviving-from-a-cyber-attack/ (2018). Accessed on 4 July 2024
3. Maritime Cyber Security. [Online]. Available from: https://www.dnv.com/maritime/insights/topics/maritime-cyber-security/ (2022). Accessed on 4 July 2024
4. Introduction cooperation on maritime cybersecurity Atlantic Council, [Online]. Available from: https://www.atlanticcouncil.org/in-depth-research-reports/report/cooperation-on-maritime-cybersecurity-introduction/ (2021). Accessed on: 4 July 2024
5. Maritime Cyber Resilience Prosjektbanken, [Online]. Available from: https://prosjektbanken.forskningsradet.no/project/FORISS/295077. Accessed on 4 July 2024
6. MariCybERA, [Online]. Available from: https://maricybera.taltech.ee/ (2021). Accessed on: 4 July 2024
7. Maritime Cyber Threats research group University of Plymouth, [Online]. Available from: https://www.plymouth.ac.uk/research/maritime-cyber-threats-research-group (2016). Accessed on: 4 July 2024
8. International Maritime Organization, MSC-FAL.1-Circ.3 - Guidelines on maritime cyber risk management (2017). Accessed: 15 Nov 2020
9. International Maritime Organization, MSC-FAL.1/Circ.3/Rev.2 - Guidelines on maritime cyber risk management (2022). Accessed: 04 Nov 2024
10. B. Svilicic, J. Kamahara, M. Rooks, Y. Yano, Maritime cyber risk management: an experimental ship assessment. J. Navigat. **72**(5), 1108–1120 (2019). [Online]. Available: https://doi.org/10.1017/S0373463318001157
11. International Maritime Organization, Resolution MSC.428(98) - Maritime cyber risk management in safety management systems (2017). https://wwwcdn.imo.org/localresources/en/KnowledgeCentre/IndexofIMOResolutions/MSCResolutions/MSC.428(98).pdf
12. AIS for Safety and Tracking: A Brief History - Global Fishing Watch, [Online]. Available from: https://globalfishingwatch.org/article/ais-brief-history/. Accessed on 20 Sept 2024
13. IALA GUIDELINE - An overview of AIS, [Online]. Available from: https://www.navcen.uscg.gov/sites/default/files/pdf/IALA_Guideline_1082_An_Overview_of_AIS.pdf. Accessed on 20 Sept 2024
14. A. Androjna, M. Perkovič, I. Pavic, J. Mišković, AIS data vulnerability indicated by a spoofing case-study. Appl. Sci. **11**(11), 5015 (2021). [Online]. Available: https://doi.org/10.3390/app11115015
15. A. Androjna, I. Pavič, L. Gucma, P. Vidmar, M. Perković, AIS data manipulation in the illicit global oil trade. J. Marine Sci. Eng. **12**(1), 6 (2023). [Online]. Available: https://doi.org/10.3390/jmse12010006
16. Above us only stars - C4ADS, https://c4ads.org/reports/above-us-only-stars/. Accessed on 20 Sept 2024
17. D.M. Valentine, Now you see me, now you don't: Vanishing vessels along Argentina's waters, Technical Report (2021). [Online]. Available: https://zenodo.org/record/4893397
18. B. Ristic, B. La Scala, M. Morelande, N. Gordon, Statistical analysis of motion patterns in AIS data: anomaly detection and motion prediction, in *2008 11th International Conference on Information Fusion* (2008), pp. 1–7

19. M. Hadzagic, A.-L. Jousselme, Contextual anomalous destination detection for maritime surveillance, in *Proceedings of the Maritime Knowledge Discovery and Anomaly Detection Workshop.(July 5–6, 2016)*. Ed. by M. Vespe, F. Mazzarella. JRC Conference and Workshop Reports. ISPRA (2016), pp. 62–65
20. H.Y. Shahir, U. Glässer, N. Nalbandyan, H. Wehn, Maritime situation analysis: A multi-vessel interaction and anomaly detection framework, in *2014 IEEE Joint Intelligence and Security Informatics Conference* (2014), pp. 192–199
21. A. Amro, A. Oruc, V. Gkioulos, S. Katsikas, Navigation data anomaly analysis and detection. Information **13**(3), 104 (2022). [Online]. Available: https://doi.org/10.3390/info13030104
22. C. Iphar, C. Ray, A. Napoli, Data integrity assessment for maritime anomaly detection.Expert Syst. Appl. **147**, 113219 (2020). [Online]. Available: https://www.sciencedirect.com/science/article/pii/S0957417420300452
23. J.N. Campbell, A.W. Isenor, M.D. Ferreira, Detection of invalid AIS messages using machine learning techniques. Procedia Comput. Sci. **205**, 229–238 (2022). 2022 International Conference on Military Communication and Information Systems (ICMCIS). [Online]. Available: https://www.sciencedirect.com/science/article/pii/S1877050922008894
24. C. Gamage, R. Dinalankarac, J. Samarabandu, et al., A comprehensive survey on the applications of machine learning techniques on maritime surveillance to detect abnormal maritime vessel behaviors. WMU J Marit Aff. **22**, 447–477 (2023)
25. S. Muggleton, Inductive logic programming. New Gener. Comput. **8**, 295–318 (1991)
26. A. Cropper, S. Dumančić, R. Evans, S.H. Muggleton, Inductive logic programming at 30. Mach. Learn. **111**(1), 147–172 (2022)
27. T. Eriksen, G. Høye, B. Narheim, B. Jensløkken Meland, Maritime traffic monitoring using a space-based AIS receiver. Acta Astronautica **58**(10), 537–549 (2006). [Online]. Available: https://www.sciencedirect.com/science/article/pii/S0094576506000233
28. Y. Chen, Satellite-based AIS and its comparison with LRIT. TransNav Int. J. Marine Navigat. Safety Sea Transport. **8**(2), 183–187 (2014). [Online]. Available: https://doi.org/10.12716/1001.08.02.02
29. International Maritime Organization, *Solas: Consolidated Text of the International Convention for the Safety of Life at Sea, 1974, and Its Protocol of 1988, Articles, Annexes and Certificates, Incorporating All Amendments in Effect from 1 January 2020*, ser. IMO Publication (2020). [Online]. Available: https://books.google.hu/books?id=JKULzgEACAAJ
30. International Maritime Organization, Resolution A.917(22) Guidelines for the Onboard Operational Use of Shipborne Automatic Identification Systems (AIS) (2001)
31. F. Cabrera, N. Molina, M. Tichavska, V. Araña, Automatic identification system modular receiver for academic purposes. Radio Sci. **51**(7), 1038–1047 (2016). [Online]. Available: https://doi.org/10.1002/2015RS005895
32. C. Hocquette, A. Cropper, Relational program synthesis with numerical reasoning. Proc. AAAI Conf. Artif. Intell. **37**(5), 6425–6433 (2023). [Online]. Available: https://ojs.aaai.org/index.php/AAAI/article/view/25790
33. M. Balduzzi, A. Pasta, K. Wilhoit, A security evaluation of AIS automated identification system, in *Proceedings of the 30th Annual Computer Security Applications Conference*. Ser. ACSAC '14 (Association for Computing Machinery, New York, 2014), pp. 436–445. [Online]. Available: https://doi.org/10.1145/2664243.2664257
34. M. Balduzzi, AIS exposed understanding vulnerabilities & attacks 2.0 (2014)
35. B. Ristic, B. La Scala, M. Morelande, N. Gordon, Statistical analysis of motion patterns in AIS data: Anomaly detection and motion prediction, in *2008 11th International Conference on Information Fusion* (2008), pp. 1–7
36. K. Kowalska, L. Peel, Maritime anomaly detection using Gaussian Process active learning, in *2012 15th International Conference on Information Fusion* (2012), pp. 1164–1171
37. M. Vespe, I. Visentini, K. Bryan, P. Braca, Unsupervised learning of maritime traffic patterns for anomaly detection, in *9th IET Data Fusion & Target Tracking Conference (DF&TT 2012): Algorithms & Applications* (2012), pp. 1–5

38. F. Katsilieris, P. Braca, S. Coraluppi, Detection of malicious AIS position spoofing by exploiting radar information, in *Proceedings of the 16th International Conference on Information Fusion* (2013)
39. J. Coleman, Behavioral model anomaly detection in Automatic Identification System (AIS), Master Thesis, The University of Tennessee at Chattanooga, Chattanooga, 2020
40. I. Kontopoulos, G. Spiliopoulos, D. Zissis, K. Chatzikokolakis, A. Artikis, Countering real-time stream poisoning: An architecture for detecting vessel spoofing in streams of AIS data, in *2018 IEEE 16th Intl Conf on Dependable, Autonomic and Secure Computing, 16th Intl Conf on Pervasive Intelligence and Computing, 4th Intl Conf on Big Data Intelligence and Computing and Cyber Science and Technology Congress(DASC/PiCom/DataCom/CyberSciTech)*, IEEE (2018). [Online]. Available: https://doi.org/10.1109/DASC/PiCom/DataCom/CyberSciTec.2018.00139
41. A. Kullberg, I. Skog, G. Hendeby, Learning motion patterns in AIS data and detecting anomalous vessel behavior, in *2021 IEEE 24th International Conference on Information Fusion (FUSION)* (2021), pp. 1–8
42. E. d'Afflisio, P. Braca, P. Willett, Malicious AIS spoofing and abnormal stealth deviations: a comprehensive statistical framework for maritime anomaly detection. IEEE Trans. Aerospace Electr. Syst. **57**(4), 2093–2108 (2021)
43. R.O. Lane, D.A. Nevell, S.D. Hayward, T.W. Beaney, Maritime anomaly detection and threat assessment, in *2010 13th International Conference on Information Fusion* (2010), pp. 1–8
44. D. Blauwkamp, T. Nguyen, G. Xie, *Toward a Deep Learning Approach to Behavior-Based AIS Traffic Anomaly Detection* (ACM, San Juan, 2018)
45. M. Louart, J.-J. Szkolnik, A.-O. Boudraa, J.-C. Le Lann, F. Le Roy, Detection of AIS messages falsifications and spoofing by checking messages compliance with TDMA protocol. Digital Signal Process. **136**, 103983 (2023). [Online]. Available: https://www.sciencedirect.com/science/article/pii/S1051200423000787
46. M. Louart, J.-J. Szkolnik, A.-O. Boudraa, J.-C. Le Lann, F. Le Roy, An approach to detect identity spoofing in AIS messages. Expert Syst. Appl. **252**, 124257 (2024). [Online]. Available: https://www.sciencedirect.com/science/article/pii/S0957417424011230
47. MarineTraffic: Global Ship Tracking Intelligence I AIS Marine Traffic, [Online]. Available from: https://www.marinetraffic.com/. Accessed on 14 Jan 2024
48. Marine AIS data - Maritime AIS vessel tracking solutions, https://spire.com/maritime/. Accessed on 18 March 2024
49. Free AIS vessel tracking I AIS data exchange I JSON/XML ship positions, [Online]. Available from: https://www.aishub.net/. Accessed on 14 Jan 2024
50. AIS Dispatcher free AIS data sharing tool AISHub, [Online]. Available from: https://www.aishub.net/ais-dispatcher (2017). Accessed on: 14Jan 2024
51. G. Visky, A. Šiganov, U.R. Muaan, R. Varandi, H. Bahsi, L. Tsiopoulos, MarCyb dataset (2024). [Online]. Available: https://data.taltech.ee/doi/10.48726/00fa9-5xv20
52. G. Visky, A. Rohl, S. Katsikas, O. Maennel, AIS data analysis: Reality in the sea of echos, in *2024 IEEE 49th Conference on Local Computer Networks (LCN)* (2024), pp. 1–7
53. A. Cropper, R. Morel, Learning programs by learning from failures. Mach. Learn. **110**(4), 801–856 (2021). [Online]. Available: https://link.springer.com/10.1007/s10994-020-05934-z

Open Access This chapter is licensed under the terms of the Creative Commons Attribution-NonCommercial-NoDerivatives 4.0 International License (http://creativecommons.org/licenses/by-nc-nd/4.0/), which permits any noncommercial use, sharing, distribution and reproduction in any medium or format, as long as you give appropriate credit to the original author(s) and the source, provide a link to the Creative Commons license and indicate if you modified the licensed material. You do not have permission under this license to share adapted material derived from this chapter or parts of it.

The images or other third party material in this chapter are included in the chapter's Creative Commons license, unless indicated otherwise in a credit line to the material. If material is not included in the chapter's Creative Commons license and your intended use is not permitted by statutory regulation or exceeds the permitted use, you will need to obtain permission directly from the copyright holder.

Technical Considerations for Open-Source Intrusion Detection System Integration in Marine Vehicles

Gabor Visky, Dariana Khisteva, and Olaf Maennel

1 Introduction

1.1 Intrusion Detection Systems

Intrusions usually cause anomalous behaviour of their victims [1]. Anomalous behaviours can also be attributed to random system failure and unforeseen external factors. Intrusion detection systems are designed to identify anomalies attributed to cybersecurity incidents [2]. An IDS identifies and logs predefined activities in network parameters, system configurations, or user behaviours and, if programmed, can also notify staff members to investigate specific alerts and take further actions.

We can distinguish IDSs according to the source of the acquired input samples in the environment they monitor. An NIDS, serving as the first line of defence, monitors network activity, collects network-related data, and identifies malicious traffic. It scrutinises all network traffic and flags any suspicious patterns.

Conversely, a HIDS provides a deeper, more localised level of security analysis by focusing on detecting potential attacks on individual computers where the IDS is installed. This system monitors system parameters such as memory content and usage, CPU load, network traffic, processes, and user actions.

These devices, using different anomaly detection methods, rule violation detection, or signatures recognition, identify possible hostile activities [3, 4].

G. Visky (✉) · D. Khisteva
Department of Computer Science, Tallinn University of Technology, Tallinn, Estonia
e-mail: gabor.visky@taltech.ee; dakhis@taltech.ee

O. Maennel
School of Computer & Mathematical Sciences, University of Adelaide, Adelaide, SA, Australia
e-mail: olaf.maennel@adelaide.edu.au

1.2 Cyber-Situation in the Maritime Sector

Modern society relies significantly on marine transportation, which handles approximately 80% of global trade [5]. To enhance efficiency, maritime systems have become increasingly digitalised and interconnected over recent decades, introducing significant cybersecurity concerns in the sector [6–9].

Many actors in the field, companies [10, 11], universities, and research institutions [12, 13] make serious efforts to increase the cyber-resilience of the sector. However, many ships rely on obsolete technologies produced without cybersecurity considerations, making them vulnerable to cyberattacks. Because of their complexity, ships' navigational and control systems cannot be upgraded overnight, and they can only be extended with security solutions after a deep analysis of the impacts. So, vendors producing such systems are hesitant about this question regarding old products.

Cybersecurity companies provide commercial-of-the-self (COTS) solutions for various sectors focusing on IT. There is a large community of commercial and/or open-source vendors, so COTS defends against "up-to-date attackers" on the IT side on moderated costs.

Since IT is easily available, OT has also started to use it: OT components are integrated into the IT world, making OT targeted by malicious actors.

Most of these COTS products handle only IT networks, but they omit the unique needs of a ship: Specialised offerings for waterborne vessels remain limited since the attack surface in the maritime environment contains not only IT but navigational, surveillance, OT, and Industrial Control Systems (ICS) [14].

These particular characteristics are still an open issue and must be addressed: Fine-tuned, tailor-made solutions are needed. Numerous studies have explored the challenges of introducing defence measures on a ship and suggested the deployment of open-source IDSs onboard.

Jacq et al. discuss the concept of naval systems' situational awareness and introduce how to detect cyberattacks on board ships in real time and elaborate on cyber-situational awareness. They found that the host-based IDS (HIDS) cannot be installed on a computer in the ship control system without causing a warranty disruption. They offered network-based IDS (NIDS) as a feasible extension [15].

Amro et al. in [16] proposed a systematic approach for navigational message analysis to detect sensor data anomalies caused by malicious activities. They demonstrated their detection capabilities by specification-based and frequency-based detection. They also propose that NIDS be added to the networks to monitor network traffic and detect anomalies.

Visky et al. highlighted the open-source IDS integration as a response to the challenge and identified the need for a concept as a research gap. Our current study addresses this gap and introduces a concept to fertilise the development and integration of such a system into marine vehicles [17].

These publications introduce the need and applicability of open-source IDSs on ships but do not evaluate all the technical details that come up during this process. Our research widens the focus and introduces more technical details that could be evaluated.

2 Related Work

Reach IDS-related literature is available. Gupta et al. surveyed and introduced it in [18]. The study introduces 113 research articles related to IDS, intrusion prevention systems (IPSs), and intrusion detection and response systems (IDRSs). The review focuses on the literature and introduces the topic but does not share detailed considerations for marine-related IDSs.

Alkasassbeh et al. introduced IDS's state of the art [2]. The paper overviews the field in detail, mainly from a technology point of view, without focusing on the needs of a particular industry segment, like shipping, where the OT-related details are significant.

Schell et al. in [19] discuss IDS concepts for intra-vehicle communication. Their publication focuses on wireless communications (Bluetooth, WiFi) and services, like Global System for Mobile Communication (GSM), and introduces the requirements for anomaly handling. These details are limited in the case of the current ships, but they gain importance as autonomous shipping comes into the picture.

Agrawal et al. in [20] introduce the concept of federated learning, which is a decentralised learning technique. It helps preserve the privacy of what gets jeopardised since IDSs often process, store, and communicate private data. The paper introduces besides the current challenges—such as communication overhead, vulnerabilities in intrusion detection setup, and federated poisoning attacks—as well as future challenges like edge computing, implementation, and optimisation issues.

Our research, motivated by the lack of a comprehensive study, collects the considerations needed for IDS development for water surface vehicles.

3 Main Considerations and Components

Integrating an IDS into marine vehicles involves several critical considerations and components that are introduced in this section.

3.1 System Requirements and Objectives

3.1.1 Security Goals

To set up the optimal IDS for a marine vehicle, the security goals should be defined. The CIA triad, comprising confidentiality, integrity, and availability, has served as a foundational framework for computer security for several decades [21]. The STRIDE (Spoofing, Tampering, Repudiation, Information disclosure, Denial of service, Elevation of privilege) methodology, as defined by Shostack [22], includes three additional elements: authentication, non-repudiation, and authorisation [23].

Based on this attack modelling method, the main security goals can be defined. Spoofing involves the ability of an adversary to masquerade as someone or something else. Denial of service refers to compromises to the system's availability by consuming the necessary resources for its proper operation. Elevation of privilege is when an adversary can execute unauthorised actions. An IDS can identify such attacks.

Tampering refers to modifying or disrupting a system's disk, network, or memory. Repudiation relates to threats where someone denies having taken specific actions that impact the system's operation or disclaims responsibility for the resulting outcomes. Information disclosure is another threat that exposes confidential information to unauthorised individuals. An IDS cannot identify such kinds of attacks.

3.1.2 Performance Requirements

To reduce the damage caused by a cyberattack, it is crucial to identify it as soon as possible and take immediate actions. The potential risks of a cyberattack are significant, and early detection is key to mitigating these risks.

An IDS can indicate malicious activity in a network that increases the cybersecurity significantly, but it cannot implement any security measure, while an IPS can. Installation of an IPS sounds like a legitimate solution, but according to our studies, the first objective of the shipping is to safely complete the mission so that a cyber-incident cannot risk it. For example, an IPS cannot exclude a mission-critical host. This decision may be made by a cyber-expert. Since there is limited or no cyber-expert in the crew, the decision should be made in the shore control centre, based on anomaly-related information collected on the ship. The limited communication bandwidth makes this process difficult, setting a special requirement for the IDS on the board. It must indicate the anomaly in its early stage and assist its mitigation on no or limited communication with the shore.

3.1.3 Regulation Compliance

An IDS must comply with the regulations and national and international law.

Cybersecurity in maritime has been gaining momentum because of recent incidents [24]. To handle this emerging problem, authorities are making significant efforts. To support effective cyber-risk management and safeguard shipping from cyber-threats and vulnerabilities in 2017, IMO issued a guideline containing high-level recommendations on maritime cyber-risk management [25, 26]. However, guidelines are just recommendatory.

To encourage administrations to appropriately address cyber-risks in existing safety management systems, no later than the first annual verification of the company's Document of Compliance after 1 January 2021 [27] in June 2017, the Maritime Safety Committee also adopted the resolution for Maritime Cyber-Risk Management in Safety Management Systems.

International Ship and Port Facility Security (ISPS) Code established by the IMO partly regulates the cybersecurity risk assessment [28]. These regularities do not specify particular characteristics of cybersecurity solutions on a ship, so there are no precise requirements that an IDS should meet.

According to the data protection laws, organisations must ensure that their use of an IDS complies with relevant data protection laws, such as the General Data Protection Regulation (GDPR) [29] in the European Union, the California Consumer Privacy Act (CCPA) [30] in the United States, or other applicable regulations. These laws typically require transparency in data collection, the right of individuals to access their data, and strict controls on data processing.

In some jurisdictions, organisations may be required to inform employees or network users that their activities are being monitored. In certain cases, explicit consent may be necessary before implementing an IDS, especially if it involves monitoring personal communications.

3.2 Selection of the IDS

After analysing the literature, experts' opinions, and the current situation, we found that installing an NIDS into ships' networks is optimal for increasing their resilience with anomaly early detection.

The market offers several commercial NIDSs [31–34]. These products have high capacity and scalability, which are not needed on ships because of their limited network size, but they cannot be extended with ship-specific features. The deployment of an open-source NIDS, known for its reliability, can effectively overcome these limitations. To find the optimal one, many details should be considered.

Adeeb Alhomoud et al. compared Suricata and Snort in high-speed networks. According to their results, the virtualisation significantly increases the number of packet drops. Additionally, Suricata performed better on Linux, while Snort on FreeBSD, especially when handling high speeds [35].

Traditional IDSs can detect known attacks by predefined rules or anomaly detection using baselines. Modern technologies collect and analyse vast amounts of data. To increase the efficiency of these IDSs, the most relevant features should be selected to shrink the dataset.

Harbola et al. discuss four kinds of feature selection methods: filter-, wrapper-, embedded methods and classification algorithm [36]. Gül and Adali [37] analysed the features of the NSL-KDD (Network Security Lab—Knowledge Discovery in Databases) dataset [38]. According to their studies, the number of the selected features depends on the attack types we want to detect. They found the most important ones in the *dst_bytes* and the *count features*. In the second-degree service, logged_in, root_shell, srv_diff_host_rate and dst_host_count were the most important.

The long-life conditions are also important when choosing an IDS. The support of the different communal products often ends within a few years, unlike in the case of open-source solutions. In their case, the activity of the community defines the "quality of the support".

3.3 Architecture

3.3.1 System Architecture

There can be several networks on the ships, which have different impacts on the missions. For example the wireless network providing Internet access to the passengers is not that mission critical like the navigation or propulsion network. While designing the architecture, this fact should be evaluated.

There are different approaches to keep these networks isolated.

Every network can have a separate, low-profile IDS, and the alerts and logs can be aggregated. This solution has moderated deployment and high maintenance costs since every IDS must be updated separately. At the same time, the different IDSs can be highly trained according to the characteristics of the given network, which can improve their sensitivity and moderate the false positive rate. This solution can also handle special typologies, such as link topology.

Another solution is the deployment of separated sensors into the different networks, and all of them send the prepossessed and unified data to a central IDS. This solution can still handle the different characteristics of the networks but needs a higher-performance central IDS.

The third solution, a high-performance centralised IDS that processes all the network traffic from the different networks, is a considerable option. However, it is important to note that this solution cannot guarantee the isolation of the networks, and the unique features of the networks are crucial from an anomaly detection perspective.

3.3.2 Network Topology

A ship may have multiple distinct networks, each serving a specific function. Visky et al. in [39] identified the following networks on a passenger ferry:

- **The administrative network**, a tree topology structure on the ship that is connected to the company's virtual private network (VPN), serves multiple functions. It ensures continuous communication with the headquarters, supports administrative tasks such as status reporting and map update downloads, and offers a monitored Internet connection for administrative use.
  The network is consistently connected via WiFi at ports, and during voyages, it connects over 4G or 5G mobile networks whenever the service is available. Satellite communication is not utilised for this purpose.
- **The navigational network** is a partially isolated system that connects navigation-related devices such as Electronic Chart Display and Information Systems (ECDIS), Integrated Navigation Systems (INSs), Multifunctional Displays (MFDs), Data Collector Units (DCUs), and RADAR. This network operates with a redundant ring topology and receives propulsion-related data from the Propulsion Control Network through a one-way connection.
- **The propulsion control network** is a partially isolated, redundant system that facilitates communication between the bridge and the propulsion automation. The passenger ferry's propulsion can be operated in fully manual mode, allowing control of propulsion and direction via physical switches in the engine room during emergencies. This network transmits propulsion-related data to the Navigational Network through a one-way connection.
- **The Cargo Handling Network** is an isolated system of wired and wireless devices. It supports cargo handling and administration activities. Cargo-related data sent from the shore in a specific format is imported into the offline cargo management system.
- **The public WiFi network** is an isolated system specifically designed to provide Internet access to passengers through a dedicated WiFi infrastructure. This network ensures a seamless online experience for passengers while onboard. It incorporates advanced network devices that manage user connections, ensuring logical separation to maintain security and privacy.
- **The network for the independent support company** is an isolated system that comprises both wired and wireless connections. It supports the ferry's onboard restaurant operations, facilitating essential functions such as order management, payment processing, and other related services. Being isolated from other networks provides enhanced security and reliability, ensuring that the restaurant can operate independently without interference from other onboard systems.

It is highly recommended that these networks be physically isolated to maintain the highest security standards. If this is the case, each network will need to be monitored individually, making the installation of an IDS more complex.

To adopt the IDS to the above-mentioned requirements, network-related data should be collected in every isolated network and sent to the centralised IDS to ensure comprehensive coverage.

3.4 Sensor Placement

Sensor placement is essential during IDS planning. We conducted a literature survey to identify the most prominent approaches.

According to Noel and Jajodia the optimal sensor placement can be determined by an attack graph analysis [40]. During this procedure, the critical assets and paths through the network can be highlighted along with the prioritisation of alerts and effective attack response.

Chen et al. in [41] evaluated the IDS sensor placement to find an optimal trade-off. According to their results, the optimal placement of sensors depends on our main objective. It may be influenced by the attack type, we wish to detect, the price, or the placement of firewalls and servers.

Based on the literature survey, there is no common method for determining the optimal sensor placement. It requires deep knowledge about the actual networks and their topology, as well as the needs and main objectives.

In order to avoid single points of failure and ensure reliable detection, it is crucial to deploy redundant sensors. It involves the use of multiple sensors to monitor the same parameter or condition in different network segments.

This approach increases reliability by making the IDS more failure-resistant: If one sensor fails, others can continue to provide the necessary data, ensuring continuous monitoring and detection. Besides that, it helps with error detection. Redundant sensors can support the cross-validation of the data and indicate a malfunction in one of the sensors.

3.5 Detection Mechanisms

3.5.1 Signature-Based Detection

One of the two primary methods an IDS uses to detect intrusion is the signature-based or misuse detection. This type of IDS is particularly effective because most attacks have unique signatures that can be identified. The simplicity and practicality of this method lie in its straightforward process learning to specify a pattern and deciding which patterns should trigger the IDS.

Signature-based IDSs operate using rule sets, which are files containing a defined traffic pattern that triggers a specified reaction. These rules are crucial as they can contain a variety of patterns, from previous attack sequences to known system vulnerabilities, that can trigger the IDS's reaction.

Each intrusion triggers unique reactions, such as denied access to a file or directory, failed login attempts, failed attempts to run an application, etc. These unique patterns are then used to detect and alert when a similar attack is happening in the future. Data is collected from different places, such as network traffic, particular resource usage, the number of requests from the same IP, etc. After this data is collected and analysed, it could be added to the knowledge base for future uses.

Some advantages of such an IDS include easier detection of attacks if the signatures are well known. In modern systems, pattern matching is optimised, making it effective and quick. Such systems are versatile, and rules might be easier to understand compared to anomaly baselines. On the other hand, the collection of signatures must be constantly updated. It is possible that a signature-based IDS fails to identify unique attacks. Rules may be redundant and use up computational resources to calculate an already evaluated result in a different way [42, 43].

3.5.2 Anomaly-Based Detection

The other section of IDSs based on how they detect intrusions is the anomaly-based IDSs. Their techniques are fundamentally different from signature-based IDSs, so their usage differs greatly. Their perspective is opposite to the signature-based variants. They do not monitor the individual packets one by one, but instead, they compare their behaviour to a standard pattern, and if that differs, it gets classified as an anomaly.

Unlike signature-based versions, anomaly-based IDSs do not rely on rule sets. Instead, they adapt to the system's behaviour profile, which defines normal activities. This adaptability ensures that any significant deviation from the norm triggers the necessary countermeasures, reinforcing their effectiveness.

The IDS refers to a behaviour profile that defines the normal behaviour of the system. Any variation from the normal will trigger alarms. This version can detect zero-day exploits. Anomaly detection can be done during run time or later down the line. Like AIDSs, anomaly-based IDSs are great against protocol or port missuses, detecting DoS attacks with crafted IP packets and other network or resource failures [44]. Attacks without known signatures are also detectable, like new worms or viruses.

The advantages of anomaly-based IDSs are significant. They can detect new versions of attacks that lack known fingerprints, instilling confidence in their ability to keep systems secure. As time passes, behaviour profiles can become more advanced, and custom profiles for different networks and applications can be created to detect unique system attacks. However, it might be hard to understand and make a profile. It is challenging to find the boundary between normal and abnormal behaviour. All protocols analysed must be well defined and tested for accuracy; otherwise, malicious behaviour might be associated with normal behaviour [42–44].

3.6 Data Collection, Storage, and Analysis

The sheer volume of data generated by IDSs necessitates significant storage capacity and advanced data management solutions to handle the influx efficiently. Furthermore, the urgency and importance of the work in threat detection is underscored by the need for real-time analysis. This involves sophisticated algorithms and substantial computational resources, and the need for high-speed processing capabilities and low-latency data handling. The diverse nature of data types collected, ranging from network traffic logs to application-specific information, further highlights the need for versatile analytical tools and methodologies.

Implementing robust log management practices for storing and analysing IDS logs involves several steps to ensure efficient storage, timely analysis, and comprehensive security monitoring.

To simplify data management and analysis, a central data server should be deployed to collect data from various IDS sensors across the network. The data should be collected in standardised formats (e.g., JSON, syslog) to ensure consistency and facilitate easier parsing and analysis.

Data storage should be scalable. Depending on the network's characteristics, the amount and generation speed of the data can vary. Modern, scalable storage solutions such as cloud storage or distributed file systems (e.g., Hadoop HDFS) can handle large volumes of log data, but on ships, the limited external communication capacity restricts the applicable technologies.

The volume of the data also involves data retention policies based on compliance requirements and business needs. To use the storage capacities optimally, it is worth moving older logs to less expensive storage options.

Different IDSs and other data sources may send the logs in different formats, which needs log parsing and normalisation. During this process, the logs are transferred into a common format, making them easier to analyse. The collected data should be indexed to enable quick and efficient querying.

To improve the IDS's sensitivity, log data should be enriched with contextual information such as geolocation, threat intelligence feeds, and asset data. This method can enhance data analysis and incident response.

The collected logs should be integrated with a Security Information and Event Management (SIEM) system for real-time monitoring, correlation, and analysis.

3.7 Alerting and Response

Once a potential security threat or anomaly is detected, to reduce the impact and damages an IDS should trigger alerts for the responsive system and associated security personnel.

During the critical initial detection phase, the IDS diligently monitors network traffic or system activities to identify any suspicious behaviour, such as unusual login attempts, abnormal data transfers, or known attack signatures. This is where your expertise comes into play. When a potential threat is detected, the IDS generates an alert, providing details such as the type of anomaly detected, the source and destination IP addresses, timestamps, and other relevant metadata. The IDS should correlate multiple alerts to identify patterns that indicate a larger, coordinated attack. Your role in this phase is crucial, as it reduces the likelihood of false positives and helps in understanding the full scope of the threat.

Since the response process is crucial for mitigating the impact of an intrusion, minimising damage, and protecting the organisation's assets, alerts must be escalated based on their severity. Predefined incident response protocols should be defined to minimise the delay and maximise the efficiency of the action taken.

These actions must consist of initial triage, in which a responsible personnel or automated system reviews the alert to determine its validity and assess the potential impact. This may involve automated log or network traffic analysis. If a higher level of action is needed, such as a manual log or network traffic analysis or forensic tools to gather more information, a cybersecurity analyst must be involved.

Depending on the threat, immediate actions must be taken to contain the intrusion, such as isolating affected systems, blocking malicious IP addresses, or disabling compromised accounts. Longer-term mitigation strategies might include applying patches, reconfiguring firewalls, or enhancing security policies to prevent similar incidents in the future.

After containment, efforts focus on restoring normal operations. This might involve cleaning up malware, restoring data from backups, or repairing damaged systems. These actions need deep IT knowledge and cannot be done as a remote operation. During the process—to prevent recurrence—additional controls or changes should be implemented, such as improving monitoring, updating IDS rules, or conducting staff training.

In summary, the alerting and response in an IDS is a multistep process that involves detecting threats, generating alerts, investigating and containing incidents, and finally recovering from and learning from the event. A well-designed alerting and response system is essential for maintaining the security and integrity of an organisation's network and systems.

3.8 *Testing and Validation*

The performance of the detection and alerting system should be validated through a series of comprehensive tests. These tests should encompass the entire environment and evaluate not only the accuracy of anomaly detection but also the effectiveness of the system's response.

3.8.1 Performance Analysis and Detection Testing

Several methods can be used to test and validate IDSs. Each method provides insights into different aspects of the IDS, ensuring its effectiveness across a range of scenarios.

The IDS's ability to differentiate between normal and abnormal traffic patterns is a key function in maintaining network security. This is tested through traffic analysis and anomaly detection, where normal and abnormal traffic patterns are introduced into the network to see how the IDS performs. This process can be conducted with publicly available datasets:

- **KDD Cup 99** [45]: One of the most well-known datasets for intrusion detection, which contains labelled data for different types of network attacks and normal traffic. According to Tavallaee et al., it suffers deficiencies: The dataset contains a huge number of redundant records, and they appear in the training and test sets as well. It makes the classifiers biased, causing a high classification rate [46].
- **NSL-KDD** [47]: An improved version of KDD Cup 99, which focuses on attack and normal traffic. It does not suffer from any of the mentioned problems. Furthermore, the number of records in the train and test sets is reasonable, allowing the experiments to be run on the complete set without the need to randomly select a small portion [46].
- **CICIDS 2017** [48]: The dataset contains benign and common attacks, which resemble real-world data. It includes 86 network-related features that also contain IP addresses and attack types. It also includes the results of the network traffic analysis: flows labelled based on many features, along with their definition.
- **UNSW-NB15** [49]: This dataset includes normal traffic and nine types of attacks namely, Backdoors, Fuzzers, DoS, Analysis, Generic, Exploits, Worms, Reconnaissance, and Shellcode [50].

The common characteristic of the above-mentioned datasets is that they aim to improve IDSs, focusing mainly on IT but not OT systems. Another shortcoming is the lack of maritime-related data.

Visky et al. published the MarCyb dataset that contains benign and attacks in ship OT networks [51]. Along with the common attacks, such as Address Resolution Protocol (ARP) spoofing and DoS, this dataset contains attacks against navigation systems, like Electronic Chart Display and Information System (ECDIS) and Global Positioning System (GPS). Furthermore, there are attacks against the navigation process, like man overboard flooding, collision alerts, and logically invalid data encoding.

The performance analysis result shows the rate of false positives (benign events flagged as threats) and false negatives (actual threats missed by the IDS).

3.8.2 Penetration Testing

Penetration testing is a method used to evaluate the security of a computer system, network, or web application by simulating an attack from malicious outsiders or insiders. The goal is to identify vulnerabilities that could be exploited by attackers, assess the effectiveness of the defences in place, and recommend improvements [52]. This helps identify how well the IDS detects real-world attacks.

The process begins with a crucial phase, the reconnaissance. This step is instrumental in understanding the target system and defining the scope of the test. It provides vital information about the target, such as domain names, IP addresses, network topology, and any public-facing services or applications. This comprehensive understanding allows for the creation of a detailed plan that outlines the targets, methods, and schedule for the penetration test, ensuring you are fully prepared for the task at hand.

The next step in the process is the scanning phase, a proactive measure to identify potential entry points and vulnerabilities in the target system. Automated tools like Nmap [53] and Nessus [54] can scan the network for open ports, running services, and potential vulnerabilities, such as outdated software, misconfigurations, or unpatched vulnerabilities. This vigilance is key to staying ahead of potential threats.

Exploitation of the identified vulnerabilities gives unauthorised access to the system. This process uses tools like Metasploit [55] or other cybersecurity frameworks to simulate various attack scenarios, such as SQL injections, DDoS attacks, or malware infections.

The successful attack demonstrates the system's weakness and gives the attacker the opportunity to deploy a backdoor, create hidden user accounts, or install malware to ensure continued access even if the initial vulnerability is patched. Establishing this foothold helps the attacker maintain access, monitor the system, and attempt to move laterally within the network to access more sensitive areas or data.

The activity's result is a report that includes the penetration test's findings and provides actionable recommendations. The detailed report outlines the vulnerabilities discovered, the methods used to exploit them, and the potential impact of each vulnerability.

3.9 Maintenance and Updates

Maintenance and updates for an IDS are crucial to ensuring that the system functions effectively in detecting and responding to potential threats. The tasks involved can be broken down into several key areas.

Maintaining up-to-date signature databases is essential for signature-based IDS. New threats and vulnerabilities emerge regularly, so the IDS needs regular updates to recognise these new patterns. Creating and refining custom rules specific to the organisation's environment can improve the accuracy of the IDS and reduce false positives.

System software and firmware updates include the IDS software and its underlying operating system or firmware to ensure they have the latest security patches and enhancements. This helps to protect the IDS itself from vulnerabilities. This process covers the implementation of new features and improvements provided by the IDS vendor that can enhance detection capabilities, performance, or usability.

Response procedures should also be updated based on the latest threats and the IDS's capabilities to ensure the most efficient incident response processes.

During regular maintenance, the IDS can be fine-tuned to maximise its sensitivity and minimise false positives. Revision of the IDS logs helps this process, by trend, repeated false positives, or new types of suspicious activity identification. The fine-tuning can involve the modification of thresholds, rules, or response actions.

Logs, configuration files, and other critical IDS data should be archived regularly to support incident response, disaster recovery, or forensics activity in the future. Maintaining the incident response plan, including regular testing of backup procedures, is also essential to quickly restore IDS functionality in case of failure.

Continuously monitoring the IDS to ensure it is operating correctly. This includes checking that sensors are functional, the network connection is stable, and the IDS is correctly logging and alerting. Besides that, during the maintenance, the IDS testing ensures that it correctly detects known threats. This could involve running simulations or using test signatures.

4 Conclusion and Future Work

4.1 Summary

In our literature analysis, we found results that highlight the usability of the open-source IDSs on ships, but we did not find any that introduces a concept with all the technical aspects to fertilise the development and integration of such a system into marine vehicles.

In our research we reviewed the existing literature to answer this need and collected and introduced the relevant challenges. We highlighted many aspects to be considered.

4.2 Main Findings

The ship usually needs IT or cybersecurity experts, so the alerts must be handled by nonexpert personnel, or they could be managed from the shore, which makes the situation difficult because of the limited communication bandwidth.

Since the ship has limited or no IT experts on board, the crew should clearly know if they can start or carry on the mission in the given cyber-situation or, if they cannot, how they can mitigate the problem.

The effectiveness of the IDS depends on its accuracy—the ability to detect real threats accurately while minimising false positives—and speed of detection, which can be increased by an automated response process. While it can speed up the response process, certain alerts may require human analysis to understand and respond to the threat fully.

Effective response to cyber-threats requires a well-coordinated system. This involves the IDS, other security tools (e.g., firewalls, SIEM systems), and the security team working together. Integration and communication are key to this well-organised and efficient system.

All these findings should be held in the forehead during the design of an IDS.

4.3 Limitation and Future Work

Our research was limited to the technical aspects. However, the administrative, human-related, financial, and legal aspects also need to be examined, which we plan to do in future work.

Acknowledgment This research was funded by the EU Horizon2020 project MariCybERA (agreement No 952360).

References

1. S. Hajiheidari, K. Wakil, M. Badri, N.J. Navimipour, Intrusion detection systems in the internet of things: a comprehensive investigation. Comput. Netw. **160**, 165–191 (2019). [Online]. Available: https://doi.org/10.1016/j.comnet.2019.05.014
2. M. Alkasassbeh, S. Al-Haj Baddar, Intrusion detection systems: a state-of-the-art taxonomy and survey. Arabian J. Sci. Eng. **48**(8), 10021–10064 (2022). [Online]. Available: https://doi.org/10.1007/s13369-022-07412-1
3. A.S. Ashoor, S. Gore, Importance of intrusion detection system (IDS). Int. J. Sci. Eng. Res. **2**(1), 1–4 (2011)
4. M. Pihelgas, A comparative analysis of open-source intrusion detection systems. Tallinn University of Technology & University of Tartu, Tallinn (2012)
5. H.N. Psaraftis, *The Future of Maritime Transport* (Elsevier, Amsterdam, 2021), pp. 535–539. [Online]. Available: https://doi.org/10.1016/b978-0-08-102671-7.10479-8
6. M. Afenyo, L.D. Caesar, Maritime cybersecurity threats: gaps and directions for future research. Ocean Coastal Manag. **236**, 106493 (2023). [Online]. Available: https://doi.org/10.1016/j.ocecoaman.2023.106493
7. E.P. Kechagias, G. Chatzistelios, G.A. Papadopoulos, P. Apostolou, Digital transformation of the maritime industry: a cybersecurity systemic approach. Int. J. Critical Infrast. Protect. **37**, 100526 (2022). [Online]. Available: https://www.sciencedirect.com/science/article/pii/S1874548222000166

8. V. Bolbot, K. Kulkarni, P. Brunou, O.V. Banda, M. Musharraf, Developments and research directions in maritime cybersecurity: a systematic literature review and bibliometric analysis. Int. J. Critical Infrast. Protect. **39**, 100571 (2022). [Online]. Available: https://www.sciencedirect.com/science/article/pii/S1874548222000555
9. M.A. Ben Farah, E. Ukwandu, H. Hindy, D. Brosset, M. Bures, I. Andonovic, X. Bellekens, Cyber security in the maritime industry: a systematic survey of recent advances and future trends. Information **13**(1), 22 (2022). [Online]. Available: https://www.mdpi.com/2078-2489/13/1/22
10. Maritime Cyber Security. https://www.dnv.com/maritime/insights/topics/maritime-cyber-security/ (2022). Undefined 7/4/2024
11. Introduction Cooperation on Maritime Cybersecurity Atlantic Council. https://www.atlanticcouncil.org/in-depth-research-reports/report/cooperation-on-maritime-cybersecurity-introduction/ (2021). Undefined 7/4/2024
12. Maritime Cyber Resilience Prosjektbanken. https://prosjektbanken.forskningsradet.no/project/FORISS/295077. Undefined 7/4/2024
13. Maricybera. https://maricybera.taltech.ee/ (2021). Undefined 7/4/2024
14. G. Potamos, S. Theodoulou, E. Stavrou, S. Stavrou, Maritime cyber threats detection framework: Building capabilities. in *Information Security Education - Adapting to the Fourth Industrial Revolution*, ed. by L. Drevin, N. Miloslavskaya, W.S. Leung, S. von Solms (Springer International Publishing, Cham, 2022), pp. 107–129
15. O. Jacq, D. Brosset, Y. Kermarrec, J. Simonin, Cyber attacks real time detection: towards a cyber situational awareness for naval systems, in *2019 International Conference on Cyber Situational Awareness, Data Analytics and Assessment (Cyber SA)* (IEEE, Piscataway, 2019). [Online]. Available: https://doi.org/10.1109/CyberSA.2019.8899351
16. A. Amro, A. Oruc, V. Gkioulos, S. Katsikas, Navigation data anomaly analysis and detection. Information **13**(3), 104 (2022). [Online]. Available: https://doi.org/10.3390/info13030104
17. G. Visky, D. Khisteva, R. Vaarandi, O.M. Maennel, Towards an open-source intrusion detection system integration into marine vehicles, in *2024 66th International Symposium ELMAR (ELMAR)* (2024)
18. N. Gupta, V. Jindal, P. Bedi, A survey on intrusion detection and prevention systems. SN Comput. Sci. **4**(5), 277–290 (2023)
19. O. Schell, J.P. Reinhard, M. Kneib, M. Ring, Assessment of current intrusion detection system concepts for intra-vehicle communication, in *INFORMATIK 2020* (Gesellschaft für Informatik, Bonn, 2021), pp. 875–882
20. S. Agrawal, S. Sarkar, O. Aouedi, G. Yenduri, K. Piamrat, M. Alazab, S. Bhattacharya, P.K.R. Maddikunta, T.R. Gadekallu, Federated learning for intrusion detection system: concepts, challenges and future directions. Comput. Commun. **195**, 346–361 (2022). [Online]. Available: https://www.sciencedirect.com/science/article/pii/S0140366422003516
21. M. Whitman, H. Mattord, *Principles of Information Security* (Cengage Learning, Boston, 2021). [Online]. Available: https://books.google.ee/books?id=Hwk1EAAAQBAJ
22. A. Shostack, *Threat Modeling* (Wiley, Nashville, 2014)
23. J. Meier, A. Mackman, S. Vasireddy, M. Dunner, R. Escamillaand, A.M. Satyam, *Improving Web Application Security* (Microsoft Corporation, Redmond, 2003). [Online]. Available: https://www.microsoft.com/en-us/download/confirmation.aspx?id=1330
24. K. Tam, K.D. Jones, Maritime cybersecurity policy: the scope and impact of evolving technology on international shipping. J. Cyber Policy **3**(2), 147–164 (2018). [Online]. Available: https://doi.org/10.1080/23738871.2018.1513053
25. I. M. O. (IMO), Msc-fal.1-circ.3 - guidelines on maritime cyber risk management (2017). Accessed: 15 Nov 2020
26. I. M. O. (IMO), Msc-fal.1/circ.3/rev.2 - guidelines on maritime cyber risk management (2022). Accessed: 04 Nov 2024
27. International Maritime Organization, Resolution msc.428(98) - maritime cyber risk management in safety management systems (2017). https://wwwcdn.imo.org/localresources/en/KnowledgeCentre/IndexofIMOResolutions/MSCResolutions/MSC.428(98).pdf

28. B. Svilicic, J. Kamahara, M. Rooks, Y. Yano, Maritime cyber risk management: an experimental ship assessment. J. Navigat. **72**(5), 1108–1120 (2019). [Online]. Available: https://doi.org/10.1017/S0373463318001157
29. General Data Protection Regulation (GDPR) – legal text. https://gdpr-info.eu/. Accessed on 16 Aug 2024
30. California Consumer Privacy Act (CCPA) | state of California - Department of Justice - Office of the Attorney General. https://oag.ca.gov/privacy/ccpa. Accessed on 16 Aug 2024
31. Next-Generation Firewalls - Palo Alto networks. https://www.paloaltonetworks.com/network-security/next-generation-firewall. Accessed on 21 April 2024
32. MDR Solutions & Services from Alert Logic. https://www.alertlogic.com/managed-detection-and-response/. Accessed on 21 April 2024
33. Cisco Secure Firewall - Cisco. https://www.cisco.com/site/ca/en/products/security/firewalls/index.html. Accessed on 21 April 2024
34. Apiiro | Secure Your Development and Delivery to the Cloud. https://apiiro.com/. Accessed on 21 April 2024
35. A. Alhomoud, R. Munir, J.P. Disso, I. Awan, A. Al-Dhelaan, Performance evaluation study of intrusion detection systems. Procedia Comput. Sci. **5**, 173–180 (2011). The 2nd International Conference on Ambient Systems, Networks and Technologies (ANT-2011) / The 8th International Conference on Mobile Web Information Systems (MobiWIS 2011). [Online]. Available: https://www.sciencedirect.com/science/article/pii/S1877050911003498
36. A. Harbola, J. Harbola, K.S. Vaisla, Improved intrusion detection in DDOS applying feature selection using rank; score of attributes in kdd-99 data set, in *2014 International Conference on Computational Intelligence and Communication Networks* (IEEE, Piscataway, 2014). [Online]. Available: https://doi.org/10.1109/CICN.2014.179
37. A. Gul, E. Adali, A feature selection algorithm for IDS, in *2017 International Conference on Computer Science and Engineering (UBMK)* (IEEE, Piscataway, 2017). [Online]. Available: https://doi.org/10.1109/UBMK.2017.8093538
38. NSL-KDD | Datasets | Research | Canadian Institute for Cybersecurity | UNB. https://www.unb.ca/cic/datasets/nsl.html. Accessed on 24 July 2024
39. A. Balint, R. Vaarandi, M. Pihelgas, O. Maennel, Open source intrusion detection systems' performance analysis under resource constraints, in *2017 IEEE 15th International Symposium on Intelligent Systems and Informatics (SISY)* (2024)
40. S. Noel, S. Jajodia, Attack graphs for sensor placement, alert prioritization, and attack response, in *Cyberspace Research Workshop* (2007)
41. H. Chen, J.A. Clark, S.A. Shaikh, H. Chivers, P. Nobles, Optimising IDS sensor placement, in *2010 International Conference on Availability, Reliability and Security* (2010), pp. 315–320
42. S. Jose, D. Malathi, B. Reddy, D. Jayaseeli, A survey on anomaly based host intrusion detection system. J. Phys. Conf. Ser. **1000**(1), 012049 (2018). IOP Publishing
43. R. Kumar, D. Sharma, HyINT: Signature-anomaly intrusion detection system. in *2018 9th International Conference on Computing, Communication and Networking Technologies (ICCCNT)* (2018), pp. 1–7
44. K. Karthikeyan, A. Indra, Intrusion detection tools and techniques–a survey. Int. J. Comput. Theory Eng. **2**(6), 901 (2010)
45. KDD Cup 1999 Data. https://kdd.ics.uci.edu/databases/kddcup99/kddcup99.html. Accessed on 10 Aug 2024
46. M. Tavallaee, E. Bagheri, W. Lu, A.A. Ghorbani, A detailed analysis of the KDD Cup 99 data set, in *2009 IEEE Symposium on Computational Intelligence for Security and Defense Applications* (2009), pp. 1–6
47. GitHub - HoaNP/NSL-KDD-dataset. https://github.com/HoaNP/NSL-KDD-DataSet. Accessed on 10 Aug 2024
48. CICIDS2017 Dataset | Papers with Code. https://paperswithcode.com/dataset/cicids2017. Accessed on 10 Aug 2024
49. The UNSW-NB15 Dataset | UNSW Research. https://research.unsw.edu.au/projects/unsw-nb15-dataset. Accessed on 10 Aug 2024

50. N. Moustafa, Designing an online and reliable statistical anomaly detection framework for dealing with large high-speed network traffic. Ph.D. Dissertation, The University of New South Wales (2017). [Online]. Available: https://hdl.handle.net/1959.4/58748
51. G. Visky, A. Šiganov, U.R. Muaan, R. Varandi, H. Bahsi, L. Tsiopoulos, Marcyb dataset (2024). [Online]. Available: https://data.taltech.ee/doi/10.48726/00fa9-5xv20
52. M. Bishop, About penetration testing. IEEE Secur. Privacy **5**(6), 84–87 (2007)
53. Nmap: The Network Mapper - Free Security Scanner. https://nmap.org/. Accessed on 10 Aug 2024
54. Download Tenable Nessus | Tenable®. https://www.tenable.com/downloads/nessus?loginAttempted=true. Accessed on 10 Aug 2024
55. Metasploit | Penetration Testing Software, Pen Testing Security | Metasploit. https://www.metasploit.com/. Accessed on 10 Aug 2024

Open Access This chapter is licensed under the terms of the Creative Commons Attribution-NonCommercial-NoDerivatives 4.0 International License (http://creativecommons.org/licenses/by-nc-nd/4.0/), which permits any noncommercial use, sharing, distribution and reproduction in any medium or format, as long as you give appropriate credit to the original author(s) and the source, provide a link to the Creative Commons license and indicate if you modified the licensed material. You do not have permission under this license to share adapted material derived from this chapter or parts of it.

The images or other third party material in this chapter are included in the chapter's Creative Commons license, unless indicated otherwise in a credit line to the material. If material is not included in the chapter's Creative Commons license and your intended use is not permitted by statutory regulation or exceeds the permitted use, you will need to obtain permission directly from the copyright holder.

Enhancing Cybersecurity in Marine Vessels: Integrating Artificial Neural Networks with Inertial Navigation Systems for Resilience Against GPS Cyber-Attacks

Yiğit Gülmez

1 Introduction

The maritime industry is progressively advancing towards autonomous shipping, driven by the digitalization of marine vessels. This shift towards highly digitalized ships offers numerous benefits, including enhanced connectivity, automated responses to critical situations, and autonomous control of various systems and subsystems. These advancements lead to greater operational efficiency, improved safety, and a significant reduction in human error, ultimately safeguarding human life, the environment, vessels, and onboard systems. However, the increased connectivity also introduces new vulnerabilities, making ships more susceptible to cyber threats. Historically, critical ship systems were not connected to remote control centres or the internet, which made them relatively secure. Now, with this increased integration, these systems are at a heightened risk of cyber-attacks. Unauthorized access to a ship's autonomous control system or automation system—which controls and monitors key functions like navigation, propulsion, and auxiliary systems—could result in catastrophic consequences, including the loss of control over the vessel.

The modern ships employ information technology (IT) and operational technology (OT) systems for shore-based monitoring and control, monitor events onboard, and process the information to provide an efficient and reliable operation. However, the integration of IT and OT systems raises the security and privacy concerns for the network between ship systems and network between ship and shore control centres. In response to similar challenges in other industries, numerous methods

---

Y. Gülmez (✉)
Estonian Maritime Academy, Tallinn University of Technology, Tallinn, Estonia

VTT Technical Research Centre of Finland, Tallinn University of Technology, Espoo, Finland
e-mail: yigit.gulmez@taltech.ee

have been implemented to enhance system reliability, such as various anomaly detection strategies of system parameters to detect and mitigate cyber threats. In the context of smart grids, Panthi and Das [1] utilized binary grey wolf optimization for feature selection in their intrusion detection system. Their study, based on a specified dataset, effectively improved intrusion classification, outperforming existing benchmarks. This approach from the smart grid sector offers a practical model for enhancing maritime systems' cybersecurity, suggesting its potential applicability in detecting and managing cyber threats in modern ship operations. In another study [2], a novel methodology based on system dynamics and sensitivity analysis to assess the impacts of cyber-attacks was introduced. Their research demonstrated this methodology using the IEEE 14-bus and 300-bus electric grid models, along with the Tennessee Eastman chemical process. The methodology's effectiveness, scalability, and applicability across different sectors were highlighted through experiments in various attack scenarios. Moreover, it outperformed traditional graph-theoretic and electrical centrality metrics, showcasing its robustness in identifying and quantifying the significance of control variables and the propagation impact of cyber threats. In a related study [3], the research introduced ARTINALI#, a Bayesian-based search and score technique designed to adapt intrusion detection systems for deployment on resource-limited cyber-physical systems. This approach strategically identifies critical points for system instrumentation, balancing the need for comprehensive attack coverage with the constraints on memory and runtime. The effectiveness of ARTINALI# was demonstrated through its deployment on two distinct systems: a smart meter and a smart artificial pancreas. Results showed a reduction in the number of security monitors by 64%, leading to decreases of 52% in memory usage and 69% in runtime, while maintaining over 98% effectiveness in detecting simulated attacks. This methodology enables the detection system adaptability across various systems with varying resource capacities, significantly enhancing the speed and efficiency of attack detection in safety-critical environments. In another study [4], Goetz and Humm introduced an unsupervised anomaly detection approach for cyber-physical production systems, utilizing graph neural networks. This methodology diverges from traditional techniques that rely solely on sensor data, by incorporating a broader array of system data through graph-based modelling. This allows for capturing intricate system characteristics, interactions, and correlations. Their approach uses a reconstruction-based graph neural network that not only detects but also analyses anomalies within these complex structures. Tested in a real industrial setting, their results confirmed the method's ability to effectively identify anomalies while simultaneously accounting for system-wide correlations and interactions. This represents a significant advancement in ensuring the safety and efficiency of cyber-physical production systems. In a recent study [5], researchers introduced two innovative deep learning models designed to improve network anomaly detection: the deep nested clustering autoencoder and the deep clustering hierarchical autoencoder. These models utilize a distinctive nested branch structure with dual deep autoencoders to form hierarchical latent spaces. Integrated clustering algorithms help optimize and refine the distribution of data points within these spaces, enhancing the separation between normal and abnormal behaviours and aiding the identification of significant features. The effectiveness of these models was rigorously tested across several major network intrusion datasets. Experimental results confirmed that both models achieved higher levels of accuracy compared to existing baselines and reduced processing times at the inference stage, indicating substantial advancements in the efficiency and effectiveness of network anomaly detection. Considering all the studies mentioned above [1–5], it is evident that various machine learning algorithms play a crucial role in ensuring accurate anomaly or intrusion detection within systems. Despite the differing methodologies across these studies, a key commonality is the importance of cross-verifying system values by using other system parameters.

On the other hand, the growing importance of cyber-physical systems in marine vessels, due to their increased connectivity, brings attention to the cybersecurity threats that marine vessels face or will potentially face. Although the literature includes studies analysing cyber risks in other critical facilities, there is a noticeable gap when it comes to marine vessels, which may be very vulnerable to cyber-attacks as well as other industrial systems. Among the few studies focusing on maritime cybersecurity, some have made significant contributions, such as conducting risk assessments for integrated navigation systems and voyage data recorders using failure mode effects and criticality analysis [6, 7], analysing cyber risks throughout the voyage cycle and within electronic chart display and information systems [8, 9], and developing a specialized "Cyber Risk Assessment for Ships" [10]. However, these studies are limited in number, and only a few have used simulation-based frameworks or experimental data to investigate the potential impact of cyber-attack scenarios on ships. A notable contribution in this area is from Longo et al. [11], who developed a threat model specifically for maritime systems. Their study focused on the ship's steering gear system, where they proposed and validated a new attack methodology through comprehensive experiments, proving its effectiveness. This work is crucial as it demonstrates how targeted cyber-attacks can manipulate ship operations and suggests strategies to increase the resilience of maritime control systems.

Despite the growing focus on cybersecurity within the maritime industry, there remains a noticeable gap in research addressing the specific vulnerabilities and resilience strategies for marine vessels, particularly concerning GPS-related cyber threats. The purpose of this study is to bridge this gap by exploring the integration of artificial neural networks (ANNs) with inertial navigation systems to enhance the resilience of maritime vessels against GPS cyber-attacks. This research is significant as it not only contributes to the existing body of knowledge by providing a novel approach to safeguarding maritime navigation systems but also offers practical solutions that can be implemented to protect critical maritime operations from increasingly sophisticated cyber threats.

2 Methodology

2.1 Data Gathering and Processing

In this study, operational data from a real ship over a period of 200 days, including operational parameters of the propulsion system such as engine torque, engine speed, turbocharger speed, and exhaust gas temperature, were collected.

The part of the data set which was gathered from the ship included an hourly basis at the start of each hour in local time, daily ship draft data, sea trial test reports, COSP and EOSP times, the records of route changing, and the fixed value of the propeller pitch. Apart from this data set, the ship's AIS track records, along with weather and sea state information relevant to the ship's location and its actual heading during the sample time, were gathered from the AIS database using Marine Traffic [12] records for the same period. The collected position, weather condition, and heading data were sampled at the top of every hour in GMT time. The two data set combined in a single data set, comprising 5029 unique entries has been organized into a spreadsheet, with all local timestamps converted to GMT. A chart shows that all related data in means of source, type, and frequency are illustrated in Fig. 1.

Adjustments were made to the collected data, utilizing hourly ship positions to determine the actual speed of the ships, which was then designated as the target speed. The engine's torque was employed to calculate shaft power. Data entries from times when ships were in ports, bunkering, drifting, anchored, or moving at speeds considered very slow (under 4 knots) were excluded, as propulsion system data in such conditions could be unreliable. Additionally, records indicating route changes within these specific intervals were omitted to avoid inaccuracies in calculating the ships' actual speeds. Following these adjustments, 1104 distinct entries were retained and utilized to develop artificial neural networks. Manoeuvring periods of the ship are included in this data set. Additionally, for certain analyses, periods of manoeuvring were excluded, specifically by omitting records from the time between EOSP and COSP, as well as any entries where wind speeds exceeded 22 knots.

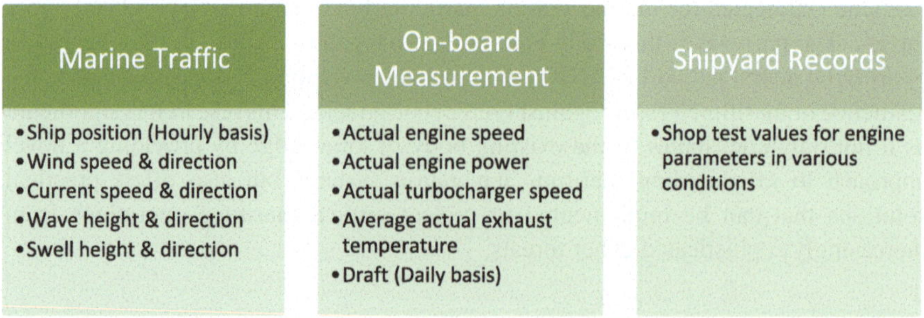

Marine Traffic	On-board Measurement	Shipyard Records
• Ship position (Hourly basis) • Wind speed & direction • Current speed & direction • Wave height & direction • Swell height & direction	• Actual engine speed • Actual engine power • Actual turbocharger speed • Average actual exhaust temperature • Draft (Daily basis)	• Shop test values for engine parameters in various conditions

Fig. 1 Data types and sources

Table 1 General information of the ship

Parameter	Value
DWT	74,999 t
GT	42,432
Service speed	15 knots
Engine power and speed	13,548 kW and 105 rpm
Propulsion	Fixed pitch, marine diesel engine
Year built	2004
LOA and Beam and draught	228 and 32.2 and 14.43 m
Net tonnage	21,821
Ballast	42,254 t
Ballast segregated	27,571 t

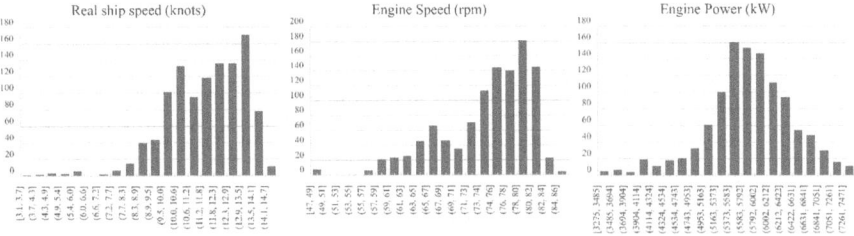

Fig. 2 Frequency distribution of key propulsion system variables in dataset

This exclusion aimed to eliminate outliers and assess how such conditions affect the model's accuracy.

The related data was collected from a crude oil carrier of which general information is illustrated in Table 1. This ship was chosen for this analysis due to its extensive navigation range, covering diverse areas such as Singapore, Brazil, the Baltic Sea, Gibraltar, and the Red Sea, which provides a wide distribution of weather conditions, as shown in Figs. 2 and 3. Additionally, the ship offers high-resolution data with minimal errors and comprehensive sea trial test records, making it particularly suitable for this type of analysis.

Figure 2 presents the distribution frequencies of key variables within the dataset, specifically focusing on actual ship speed, engine power, and engine speed. The illustration reveals a broad spectrum of values across these three parameters, indicating the dataset encompasses a variety of unique and infrequent operational states, such as reduced engine speeds, minimal loads, and both medium and low ship velocities. This diversity complicates the training process for neural networks, yet simultaneously enhances the model's ability to adapt to a wide range of scenarios.

Figure 3 shows the frequency distribution of weather and sea state variables, which is important for the study's goal to analyse how weather conditions affect changes in vessel speed relative to theoretical ship speed. The inclusion of a broad spectrum of weather and sea conditions is essential for developing an accurate neural network model. The figure reveals that the dataset includes a wide variety of weather

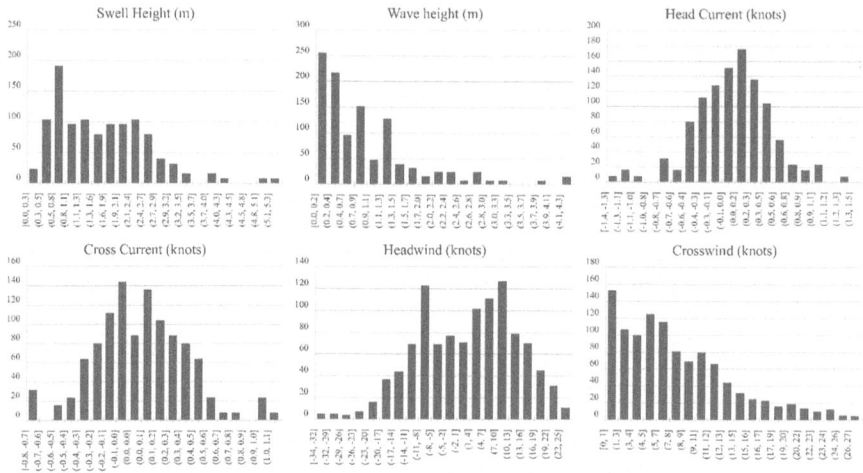

Fig. 3 Frequency distribution of weather and sea state variables in dataset

conditions, notably in wind speed, ranging from 0 to 35 knots, equivalent to 0 to 8 on the Beaufort scale. Wind direction is also considered crucial for assessing its impact on ship speed and resistance and is therefore depicted in two separate graphs for headwind and crosswind. Notably, 17.39% of the headwind data exceeds the ±15 knots range, and 9.4% surpasses ±20 knots, enhancing the model's depth. Regarding sea conditions, wave heights reach up to 4.3 m and swell heights up to 5.3 m, with wave heights exceeding 2.5 m in 6.5% of scenarios, and swell heights exceeding 3 m in 6.6% of scenarios. The dataset encompasses a wide range of wind speed, wave heights, and swell heights, demonstrating the neural network model's potential to accurately estimate conditions up to these limit values.

The dataset encompasses a wide array of essential elements, including the ship's propulsion system metrics, weather and sea conditions around the vessel, and its positional information. This study aims to explore several key research questions, such as the feasibility of predicting ship speed using data on weather, sea states, and propulsion system metrics, and the potential of determining ship speed by evaluating the influence of weather and sea states on propulsion parameters like engine power, engine speed, turbocharger speed, and exhaust gas temperature. To this end, three distinct neural network models were developed, each representing a different scenario.

In the first case, the model uses only data from the propulsion system—specifically, engine speed, engine power, turbocharger speed, and average exhaust temperature—to predict ship speed. This model assesses the effect of weather and sea state on propulsion system performance by comparing actual engine parameter values against those obtained during sea trials, which were conducted under calm sea and weather conditions. A larger discrepancy indicates a greater impact of weather and sea states on engine performance, resulting in a wider gap between theoretical and actual ship speeds. Therefore, this model aims to quantify the impact of weather and sea states on the variance between measured engine parameters and sea trial data. In the second scenario, the model incorporates the same propulsion system parameters as the first but adds wind speed and its relative direction as inputs. This decision was based on an analysis showing that wind-related parameters significantly enhance the model's accuracy in speed estimation. Hence, wind parameters were chosen as the additional inputs for improved predictive accuracy. The third scenario expands further by utilizing all available propulsion system data along with comprehensive weather and sea state information to estimate the actual speed of the ship. This comprehensive approach aims to provide a holistic view of how both propulsion dynamics and environmental conditions affect ship velocity.

The research also explored whether neural network model outputs could help establish a theoretical circle's radius around a ship's last known position, which would confine the ship's potential maximum or minimum positional changes based on its prior location. To achieve this, analyses of discrepancies were conducted to identify the extreme possible changes in position for each hour of navigation. Graphs displayed all differences between the actual ship speeds and the estimated ones derived from the neural network model for each scenario, aiding in setting the boundary values. Furthermore, conditions that significantly impact the discrepancies yet occur infrequently were isolated, leading to further discrepancy analyses with more precise data.

2.2 Artificial Neural Networks

Neural networks are an integral part of machine learning, inspired by the biological neural networks that constitute animal brains. Basically, these computational models are designed to recognize patterns and find solution to complex problems by applying the same technique used by biological neurons signal to one another [13, 14]. Neural networks include layers with interconnected nodes or neurons. Each connection represents a data pathway which transmits a signal from one neuron to another. Neural networks have three basic phases: training, validation, and testing. Determining of weights, which represents the strength of these connections, and biases, which modify the output on the activation function, are adjusted during training, allowing the network to learn from the data [15, 16]. Evaluating of the learning process and fine adjusting of weights and biases conducted in the validation process by using a dataset that is not used in learning phase. And final evaluation of the model is made in the test phase with another data set which is not seen in the learning and validation phases [16].

In the current study, the speed of a real ship was estimated by utilizing the data from the propulsion system and environmental conditions by using three alternative data combinations in three cases. The illustration of the neural network diagram for these cases is illustrated in Fig. 4.

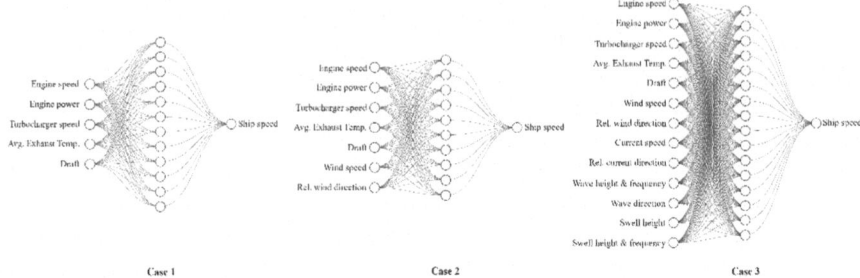

Fig. 4 Neural network diagrams for three cases

MATLAB's neural network toolbox has been used to develop the neural network models. Modelling of each case made independently by using the selected features from the same data set. Firstly, all the data set divided into three categories: training, validation, and testing with rates of 70%, 15%, and 15%, respectively. A feedforward neural network optimized by the Levenberg–Marquardt algorithm, combined with a hyperbolic tangent sigmoid activation function was used in the study to detect, map, and learn the complex nonlinear relationships between the inputs and targets. The architecture of the neural network was constructed with a single hidden layer and an output layer for all cases. The number of neurons within the hidden layers were adjusted by trial-and-error method. For this purpose, the model was run iteratively multiple times to reach the optimal configuration that resulted in the best estimation performance by means of the performance metrics used in this study. The compilation of MSE results for the top ten performing iterations for each of the three cases were assembled to demonstrate the reliability of neural networks.

All inputs and target features were standardized by the process of data preparation involved Z-score normalization as it is a crucial step for the neural network to learn effectively. After the predictions were made, a reverse process of denormalization was applied to convert the predicted values back to their original scale for a straightforward comparison with actual ship speeds. Z-score normalization was chosen through a trial-and-error process by comparing the prediction performance of the algorithms among alternative normalization methods such as min-max scaling, log transformation and square root transformation.

2.3 Calculations

The data set has been used to produce new types of data that are more appropriate to make analysis or to create neural networks study. In this study, three different neural networks models were created by using the difference of engine power from the engine power data from the sea trial tests, the difference of turbocharger speed from the turbocharger speed data from the sea trial tests, the difference of average exhaust gas temperature from the average exhaust gas temperature data from the sea trial tests, relative directions of wind, current, wave, and swell, the speed of wave and current, the height of wave and swells, the frequency of wave and swells, and the real speed of the ship. Among these variables, the height and frequency of wave and swells, and the speed of wind and current have been used directly in the model without any processing. The difference of engine power from the sea trial test was calculated with the formula below where ΔP represents the difference of engine power, subscript of "op" indicates the measurement result taken during the operation and subscript of "st" indicates the measurement result taken in same draught value during the sea trial tests. ΔRPM_{TC} and ΔT_{exh} values which indicate turbocharger speed and average exhaust gas temperature, respectively, were also calculated in a similar way.

$$\Delta P = P_{op} - P_{st} \qquad (1)$$

Relative direction of the wind, current, wave, and swell were calculated by the difference between the actual course of the ship and the direction of the phenomenon by the following formula where θ_i indicates the direction of the related parameter, θ_{course} is the course of the ship.

$$\Delta \theta_{rel} = \theta_{course} - \theta_i \qquad (2)$$

Real speed of the ship was calculated by using the Cosine rule [17] and by using the positional information of the ship recorded from the AIS data by the following formula where d is the distance travelled between the two-measurement point, r is the radius of the globe, and $\varnothing_i$ and λ_i are the latitude and longitude of point i respectively. And, the real ship speed was calculated by using the distance and certain period of time.

$$d = r \arccos\left(\sin \varnothing_a \sin \varnothing_b + \cos \varnothing_a \cos \varnothing_b \cos(\lambda_b - \lambda_a)\right) \qquad (3)$$

Theoretical speed of the ship was calculated by using the propeller pitch and engine speed by the following formula where RPM_{en} is the engine speed, υ_{th} is the theoretical ship speed, and pitch is the distance the propeller would move forward in one complete revolution if there were no slippage [18].

$$\upsilon_{th} = RPM_{en} \cdot pitch \cdot 60 \qquad (4)$$

To enhance the accuracy of the neural network model, all data utilized in the model were normalized using the Z-score method. In this method, the data is standardized to scale the attribute values for making them lie numerically in the same interval and scale to prevent the scale of the data set from affecting the importance of any parameter in the modelling process [19]. This type of normalization has a wide range of applications in neural networks studies [20–22]. The formula for calculating the Z-score of a data point is as follows where x is any value in the data set, μ is the mean value of all data points within the same feature set, and σ is the standard deviation of the data points within the same feature set.

$$Z = \frac{(X - \mu)}{\sigma} \tag{5}$$

The performance and accuracy of neural networks models were evaluated using some statistical metrics such as mean square error (MSE) and R value (R). MSE is measured as mean squared differences between output and target [23], and R is the value of the relationship between variables and quantifies the strength of the relationship between targets and outputs [24]. MSE and R were calculated as follows where n is the number of data points, y is the output value, and x is the target value.

$$\text{MSE} = \frac{1}{n} \sum_{i=1}^{n} (y_i - x_i)^2 \tag{6}$$

$$R = \frac{\sum_{i=1}^{n} (x_i - \overline{x})(y_i - \overline{y})}{\sqrt{\sum_{i=1}^{n} (x_i - \overline{x})^2 (y_i - \overline{y})^2}} \tag{7}$$

3 Findings and Discussion

The neural network approach enabled the creation of mathematical relationships between input parameters and targets, thereby improving the prediction of output values. For each case, the neural network model was evaluated using varying biases and weights to enhance the correlation between the outputs and the targets. The algorithms that achieved the highest correlation were selected as the basis for the simulation model for each case. Figure 5 displays the optimal validation performances of simulation models.

Figure 5 presents the number of epochs utilized in the model and the lowest achievable MSE for the comparison of predicted outputs and targets. The neural network modelling tool employed in this study incorporates a feature known as "early stopping", which is designed to strike a balance between providing the model sufficient attempts to learn and avoiding overfitting or performance degradation. According to the diagram, the MSE begins to rise, indicating model degradation, after a certain number of epochs, leading to the activation of the early stopping mechanism at the most advantageous point. The optimal validation performance is indicated to improve at the seventh epoch for case 1, the sixth epoch for case 2, and the sixth epoch for case 3. The corresponding MSE values are 0.0644 for case 1, 0.041055 for case 2, and 0.046508 for case 3, with lower MSE values signifying a smaller discrepancy between the model's output and the target values for each

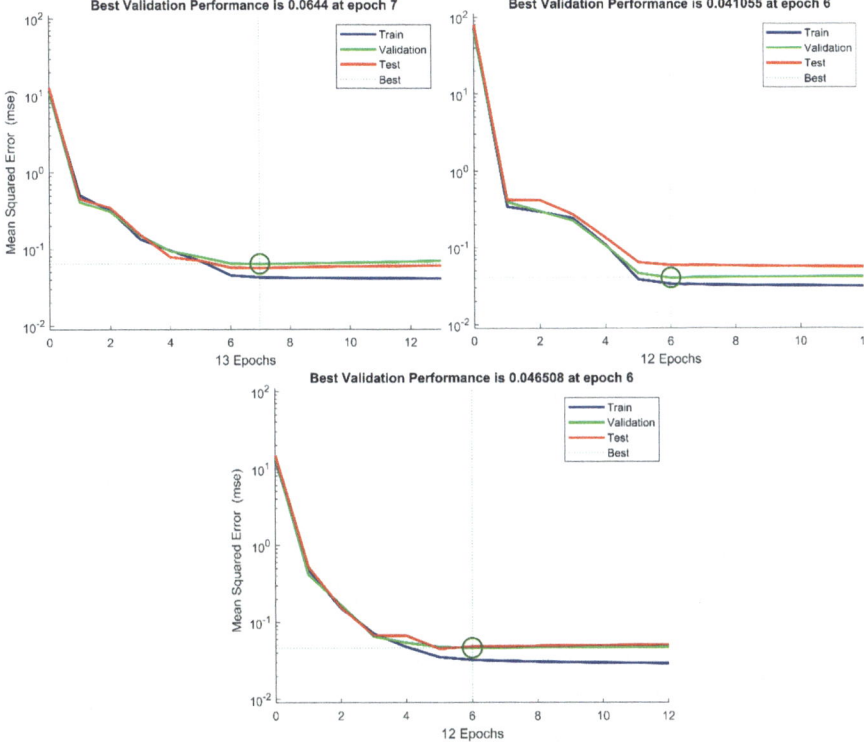

Fig. 5 MSE values and optimal epochs for three cases

scenario. These outcomes suggest a strong correlation between the simulation outputs and the targets, validating the neural network model's effectiveness in estimating the ship's speed across all three scenarios. Additionally, the results were derived from a trial-and-error process involving numerous attempts, with observations indicating generally the lowest MSE values for case 2 and the highest for case 1. The MSE values for the ten most successful trials for each case are depicted in Fig. 6.

Figure 6 displays a general correlation between MSE values and various scenarios, indicating that the most accurate simulation outcomes, or the closest approximations to the target, were achieved in case 2. In this scenario, data from certain propulsion system sensors and limited weather information, specifically wind speed and relative wind direction, were utilized to achieve the closest results to the target. Case 3 expanded on the inputs used in case 2 by incorporating additional weather parameters such as swell height, relative swell direction, swell period, wave height, relative wave direction, wave period, current speed, and relative current direction. However, despite the increased number of input parameters, including all those from case 2, case 3 resulted in higher MSE values, suggesting that the addition of more input parameters negatively affected the accuracy of the estimations.

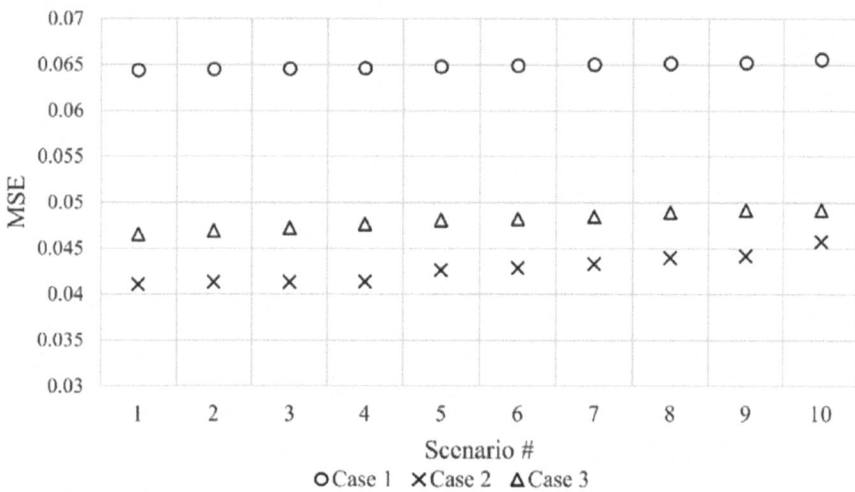

Fig. 6 Evaluation of neural network performance with best ten MSE outcomes for each case

Throughout this study, 1104 different scenarios were employed to develop the neural network model. It is possible that with an even greater variety of scenarios, case 3 could yield improved results. Observing the outcomes of case 1, where no weather-related data were used, the results were still satisfactory. This indicates that the neural network was able to infer the impact of weather conditions on propulsion system data, thereby modelling the effect of weather on ship speed without direct weather data inputs and only with propulsion system data. Despite the absence of explicit weather data in case 1, the neural network effectively modelled the influence of weather on the propulsion system, leading to highly accurate speed estimations. This outcome demonstrates the potential of using propulsion system data alone to accurately predict the impact of weather conditions on ship speed. The effectiveness of the model in estimating these impacts will be further discussed in subsequent sections of the document.

Figure 7 presents the R values for various scenarios, indicating a close alignment between targets and outputs. Notably, case 3 and case 2 exhibit comparatively higher R values, while case 1 has the lowest. Nonetheless, the R value of 0.9756 in case 1 signifies a robust relationship between targets and outputs. As shown in the tables in the figure, datasets were divided into three groups, with each group's R values calculated for every case. In every group, case 1 consistently shows the lowest R values, whereas the sequence between case 2 and case 3 fluctuates, demonstrating that R values do not follow a uniform order, especially between these two cases.

Comparing the findings extracted from Figs. 5, 6, and 7, this lack of consistent ordering between MSE and R values suggests distinct patterns of error distribution across cases. Specifically, case 3 might contain several outliers that elevate the MSE, indicating poor performance, yet the majority of predictions closely track the actual values, resulting in a high correlation coefficient. Conversely, case 2 might display

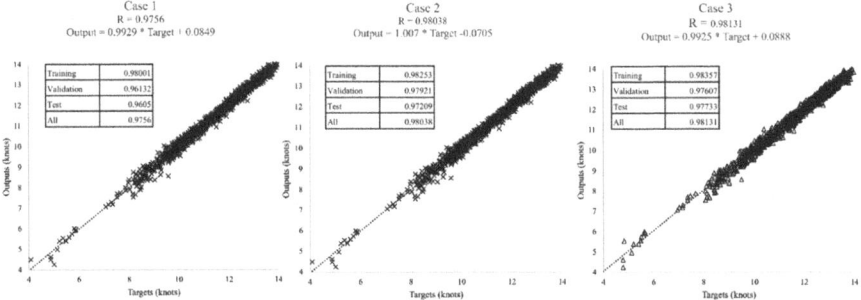

Fig. 7 Neural network models performance with R value for three cases

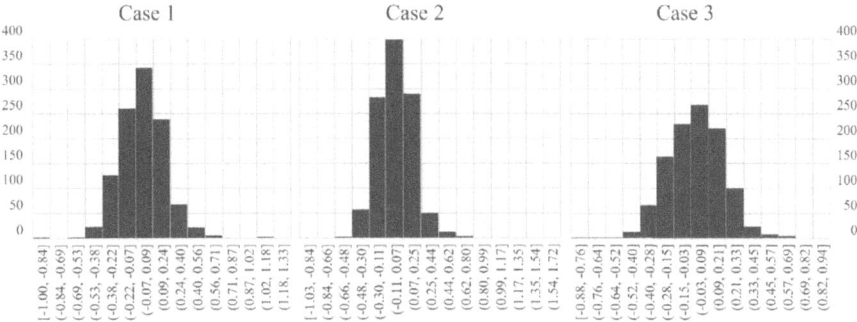

Fig. 8 Comparative discrepancy distribution assessing ship speed estimation accuracy

more evenly spread errors that, despite being smaller, do not closely mirror the trend of the actual values, leading to a reduced correlation coefficient. This implies that, in these simulation outcomes, although predictions in cases 2 and 3 are generally well-aligned, the presence of more outliers in case 3 causes its MSE to surpass that of case 2.

Figure 8 illustrates the variance in the discrepancy between targets and outputs across different scenarios. In case 1, a significant majority, 80.43%, of discrepancies fall within ±0.25 knots; this figure rises to 97.55% for discrepancies within ±0.5 knots and reaches 99.63% for those within ±1 knot. Case 2 sees a slight improvement, with 83.69% of discrepancies within ±0.25 knots, 98.55% within ±0.5 knots, and 99.72% within ±1 knot. Case 3 further refines these results, with 81.79% of discrepancies within ±0.25 knots, 98.37% within ±0.5 knots, and a perfect 100% within ±1 knot. These outcomes suggest varying levels of estimation accuracy across the three cases but consistently demonstrate a high correlation rate of discrepancies within ±0.5 knots. Based on these outcomes, it is inferred that the neural network model exhibits a high correlation in various scenarios, making it a viable alternative for estimating ship speed. Furthermore, incorporating the ship's course into the model could also enable position prediction accuracy relative to a known previous position.

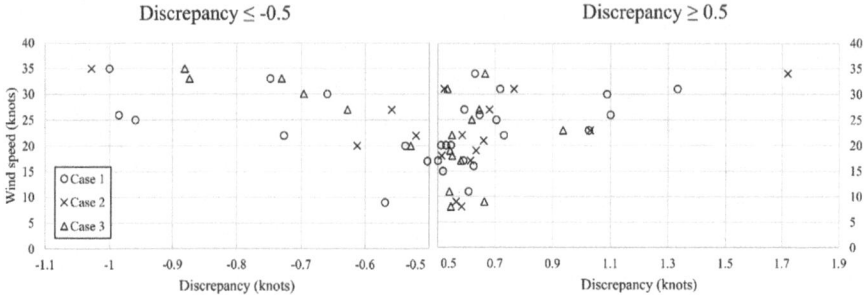

Fig. 9 Wind speed's effect on prediction accuracy beyond 0.5 knot deviations

A thorough analysis of the dataset reveals a significant correlation between specific weather conditions, notably wind speed, and the observed deviations between the actual neural network outputs and their intended targets. Highlighted in Fig. 9 are instances where the neural network's output deviations from the targets surpass 0.5 knots, with a clear indication of a relationship between these deviations and wind speed. This is particularly evident for deviations exceeding 0.66 knots, where wind speeds invariably exceed 22 knots, aligning with Beaufort scale levels of 6 or above. This pattern suggests that under conditions up to a Beaufort scale of 6, the neural network model demonstrates reliable ship speed predictions with a margin of error limited to 0.66 knots, without any exceptions. In an individual assessment of scenarios, after filtering out conditions with wind speeds over 22 knots, the model showcases a discrepancy within ±0.5 knots in 99%, 99.3%, and 99.3% of the cases for the first, second, and third scenarios, respectively. This performance translates to the model's capability to estimate ship speeds with a 99% or higher accuracy rate within a 0.5 knot threshold. Moreover, at a confidence level of 95%, the model achieves ship speed estimations within error margins of 0.38, 0.33, and 0.35 knots for the respective scenarios.

The analysis considers data involving ship movements at speeds over 4 knots, including instances of manoeuvring where ships rapidly change speed. And, it may potentially lead to data inaccuracies. To mitigate this, manoeuvring situations were removed in a separate analysis. This refined analysis, presented in Fig. 10, shows wind speed distributions and data discrepancies excluding manoeuvring scenarios. When manoeuvring periods are removed, but all weather conditions are considered, the model demonstrates a discrepancy rate within ±0.5 knots for 97.61%, 98.63%, and 98.4% of cases in the first, second, and third scenarios, respectively. Further refining the analysis by excluding weather conditions with wind speeds over 22 knots results in an even lower discrepancy rate within ±0.5 knots for 99.35%, 99.61%, and 99.61% of the cases across the three scenarios. Moreover, when both manoeuvring situations and high wind speed conditions (over 22 knots) are removed, the model achieves ship speed estimations with error margins of 0.36, 0.32, and 0.33 knots for the first, second, and third scenarios, respectively, at a 95% confidence level.

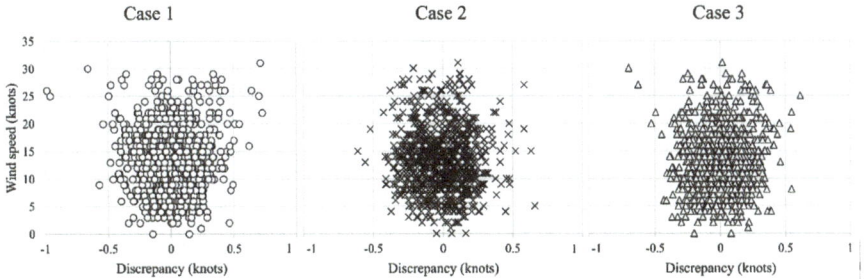

Fig. 10 Refined accuracy in ship speed predictions excluding manoeuvres and high winds

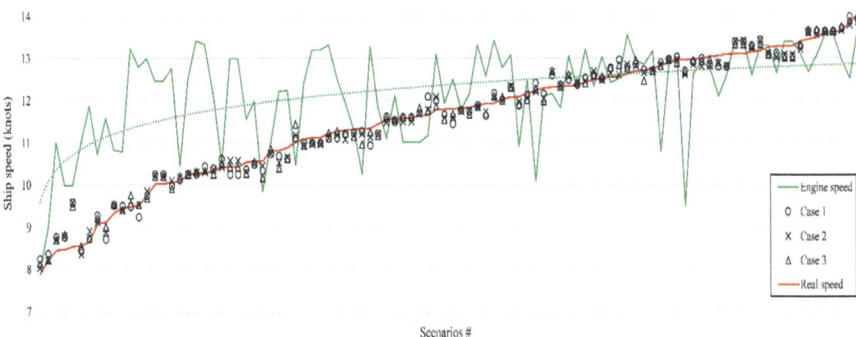

Fig. 11 Comparison of theoretical engine speed, real speed, and neural network estimations for randomly selected 100 scenarios

Figure 11 presents a graphical comparison between actual ship speed, estimated ship speeds from three different scenarios, and engine-based speed, which is derived by multiplying the propeller pitch by the propeller rpm. The comparison highlights that the neural network model provides estimates that are significantly closer to the actual ship speed compared to the engine speed estimates. This indicates an improvement in the accuracy of ship speed estimation by leveraging outputs from the propulsion system. The graph organizes the actual ship speeds in ascending order and displays both the engine speeds and the outcomes from various simulation scenarios. A selection of 100 random scenarios was included in the graph. The trendline for engine speed suggests a correlation between engine speed and actual speed, yet the mean square error between the engine speed and actual speed is considerably higher than that observed in the simulation results. These findings demonstrate the efficacy of using neural networks in yielding highly accurate approximations of actual ship speed.

After conducting thorough analysis, it becomes clear that the most favourable outcomes were observed in use case 2. Interestingly, despite the addition of numerous other input parameters in the neural network model for use case 3, which encompasses all inputs from use case 2, the level of discrepancy in use case 3 is surprisingly higher than that in use case 2. Considering that the model in use case 1 operates without utilizing any direct weather-related input parameters, its results are deemed satisfactory. Furthermore, the findings indicate that employing data from the propulsion system alongside certain weather conditions in a neural network to predict ship speed can yield highly accurate results with a 95% confidence level. In use case 1, the maximum discrepancy observed is 0.36 knots, while in use cases 2 and 3, it is reduced to 0.32 and 0.33 knots, respectively. However, these discrepancy levels are considered too significant for the method to be recognized as a highly accurate dead reckoning system. Nonetheless, these results hold importance, particularly for use case 1, as it achieves a high level of accuracy for a simulation model that does not incorporate any external information or sensor measurements from the engine room. This aspect is crucial as it could serve as an indicator of the cyber robustness of the navigation system. Given that all navigation systems on the ship's bridge are interconnected, enhancing the system's vulnerability, the approach proposed in this study could offer a potential solution for detecting cyber-attacks.

4 Conclusion

This study highlights an innovative approach of utilizing propulsion system data to accurately predict ship speed and assess the impact of weather conditions, thereby offering a critical advantage in enhancing maritime security and operational resilience. The ability to determine ship speed without relying on GPS technology is significant, especially considering the vulnerability of GPS to spoofing and jamming attacks, which pose a risk to navigational integrity. By leveraging propulsion system data, ship operators can maintain accurate speed estimations even when GPS signals are compromised, thus mitigating potential navigational errors or security breaches.

Furthermore, this method emphasizes the importance of minimizing data usage to bolster cybersecurity and provides a strategic alternative to GPS-dependent navigation. It ensures sustained operational capabilities under challenging conditions, such as weak GPS signals or deliberate interference, and enhances the ship's autonomy by relying on internal data systems. This approach simplifies data collection, enhances cybersecurity, and offers a robust method for uninterrupted, secure maritime navigation, marking a significant advancement in maritime technology with dual benefits of operational efficiency and strengthened security against cyber threats.

In addition, incorporating propulsion system data for ship speed prediction serves as a potent tool for cybersecurity and maintaining operational integrity. It plays a pivotal role in detecting cyber-attacks and identifying anomalies in GPS data. By continuously comparing ship speed estimations derived from propulsion data with GPS-reported speeds, discrepancies beyond expected environmental variances can signal potential cyber threats or GPS data manipulations. This capability is invaluable for recognizing subtle, logical changes in GPS data that might otherwise go undetected, potentially leading to significant navigational errors. Utilizing propulsion system data as an independent verification mechanism ensures the authenticity and accuracy of GPS-reported data, offering a critical safeguard against compromised GPS data due to cyber-attacks or technical malfunctions. This dual-analysis approach bolsters maritime cybersecurity by enabling early anomaly detection and facilitating timely corrective actions to ensure navigational safety and security.

Acknowledgments The research was supported by the EU Horizon2020 project MariCybERA, Agreement no. 952360.

References

1. M. Panthi, T.K. Das, Intelligent intrusion detection scheme for smart power-grid using optimized ensemble learning on selected features. Int. J. Crit. Infrastruct. Prot. **39**, 100567 (2022). https://doi.org/10.1016/j.ijcip.2022.100567
2. B. Genge, I. Kiss, P. Haller, A system dynamics approach for assessing the impact of cyber attacks on critical infrastructures. Int. J. Crit. Infrastruct. Prot. **10**, 3–17 (2015). https://doi.org/10.1016/j.ijcip.2015.04.001
3. M. Raiyat Aliabadi, M. Seltzer, M. Vahidi Asl, R. Ghavamizadeh, ARTINALI#: an efficient intrusion detection technique for resource-constrained cyber-physical systems. Int. J. Crit. Infrastruct. Prot. **33**, 100430 (2021). https://doi.org/10.1016/j.ijcip.2021.100430
4. C. Goetz, B.G. Humm, Unsupervised correlation- and interaction-aware anomaly detection for cyber-physical production systems based on graph neural networks. Proc. Comput. Sci. **232**, 2057–2071 (2024). https://doi.org/10.1016/j.procs.2024.02.028
5. V.Q. Nguyen et al., Deep clustering hierarchical latent representation for anomaly-based cyber-attack detection. Knowl. Based Syst. **301**, 112366 (2024). https://doi.org/10.1016/j.knosys.2024.112366
6. A. Oruc, A. Amro, V. Gkioulos, Assessing cyber risks of an INS using the MITRE ATT&CK framework. Sensors **22**, 8745 (2022). https://doi.org/10.3390/s22228745
7. Ö. Söner, G. Kayisoglu, P. Bolat, K. Tam, Cybersecurity risk assessment of VDR. J. Navig. **76**(1), 20–37 (2023). https://doi.org/10.1017/S0373463322000595
8. K. Shumilova, D. Shumilov, A. Maltsev, Classification of cyber risks for sea vessel's voyage cycle. Trans. Marit. Sci. **13**(1) (2024). https://doi.org/10.7225/toms.v13.n01.w20
9. G. Kayişoğlu, B. Güneş, P. Bolat, ECDIS cyber security dynamics analysis based on the fuzzy-FUCOM method. Trans. Marit. Sci. **13**(1), 20–37 (2024). https://doi.org/10.7225/toms.v13.n01.w09
10. A. Oruc, G. Kavallieratos, V. Gkioulos, S. Katsikas, Cyber risk assessment for ships (CRASH). TransNav **18**(1), 115–123 (2024). https://doi.org/10.12716/1001.18.01.10
11. G. Longo et al., Physics-aware targeted attacks against maritime industrial control systems. J. Inf. Secur. Appl. **82**, 103724 (2024, ISSN 2214-2126,). https://doi.org/10.1016/j.jisa.2024.103724
12. Marine Traffic, 2024, www.marinetraffic.com
13. P.B. Pires, J.D. Santos, I.V. Pereira, Artificial neural networks: history and state of the art, in *Encycl. Inf. Sci. Technol*, 6th edn., (2024), pp. 1–25. https://doi.org/10.4018/978-1-6684-7366-5.ch037
14. D. Sadanand, S. Bhosale, Basics of artificial neural network. Int. J. Adv. Res. Sci. Commun. Technol. **3**(3), 299–303 (2023). https://doi.org/10.48175/ijarsct-8159

15. L. Vanneschi, S. Silva, Artificial neural networks, in *Lectures on Intelligent Systems*, (Springer International Publishing, Cham, 2023), pp. 161–204. https://doi.org/10.1007/978-3-031-17922-8_7
16. Y. Gülmez, G. Özmen, Effect of exhaust backpressure on performance of a diesel engine: neural network based sensitivity analysis. Int. J. Automot. Technol. **23**, 215–223 (2022). https://doi.org/10.1007/s12239-022-0018-x
17. M. Tuomi, Bachelors thesis, SAMK—Satakunnan ammattikorkeakoulu, 2021, https://urn.fi/URN:NBN:fi:amk-202103183538
18. M. Suranto, I.M. Ariana, A. Baidowi, Analysis of the effect of pitch angle on propeller modification by considering wake distribution on propeller performance, in *IOP Conference Series: Earth and Environmental Science*, 6th International Conference on Marine Technology (SENTA 2021), 27th November 2021, Surabaya, Indonesia, vol. 972, p. 012049. https://doi.org/10.1088/1755-1315/972/1/012049
19. S.H. Javaheri, M.M. Sepehri, B. Teimourpour, Chapter 6—Response modeling in direct marketing: a data mining-based approach for target selection, in *Data Mining Applications with R*, ed. by Y. Zhao, Y. Cen, (Academic Press, 2014), pp. 153–180. https://doi.org/10.1016/B978-0-12-411511-8.00006-2
20. N. Fei, Y. Gao, Z. Lu, T. Xiang, Z-score normalization, hubness, and few-shot learning, in *Proceedings of the 2021 IEEE/CVF International Conference on Computer Vision (ICCV), Montreal, QC, Canada*, (2021), pp. 142–151. https://doi.org/10.1109/ICCV48922.2021.00021
21. G. Aksu, C.O. Güzeller, M.T. Eser, The effect of the normalization method used in different sample sizes on the success of artificial neural network model. Int. J. Assess. Tools Educ. **6**(2), 170–192 (2019). https://doi.org/10.21449/ijate.479404
22. S. Eesa, W.K. Arabo, A normalization methods for backpropagation: a comparative study. Sci. J. Univ. Zakho **5**(4), 319–323 (2017). https://doi.org/10.25271/2017.5.4.381
23. K. Tyagi, C. Rane, Harshvardhan, M. Manry, Regression analysis, in *Artificial Intelligence and Machine Learning for EDGE Computing*, (Academic Press, 2022), pp. 53–63. https://doi.org/10.1016/B978-0-12-824054-0.00007-1
24. I. Ocampo, R.R. López, S. Camacho-León, V. Nerguizian, I. Stiharu, Comparative evaluation of artificial neural networks and data analysis in predicting liposome size in a periodic disturbance micromixer. Micromachines (Basel) **12**(10), 1164 (2021). https://doi.org/10.3390/mi12101164

Open Access This chapter is licensed under the terms of the Creative Commons Attribution-NonCommercial-NoDerivatives 4.0 International License (http://creativecommons.org/licenses/by-nc-nd/4.0/), which permits any noncommercial use, sharing, distribution and reproduction in any medium or format, as long as you give appropriate credit to the original author(s) and the source, provide a link to the Creative Commons license and indicate if you modified the licensed material. You do not have permission under this license to share adapted material derived from this chapter or parts of it.

The images or other third party material in this chapter are included in the chapter's Creative Commons license, unless indicated otherwise in a credit line to the material. If material is not included in the chapter's Creative Commons license and your intended use is not permitted by statutory regulation or exceeds the permitted use, you will need to obtain permission directly from the copyright holder.

A Comprehensive Review of Social Engineering on Maritime Cybersecurity

Veera Senthil Kumar Ganesan and Mihir Chandra

1 Introduction

The shipping industry is pivotal in today's world economy. Intercontinental trade would not be possible without this industry. Shipping accounts for more than 80% of the world trade volume. Looking at today's scenario of shipping, it has become more equipped with cutting-edge technologies and digitized to meet the requirements of the growing needs of the industry. As technologies, used in shipping, are being proliferated, there are benefits such as faster communication and higher automation [1]. On the other hand, the shipping industry has to be ready to face looming challenges because of these technological advancements [2]. Cyber threats to shipping have become one of the major challenges with the increase in automation.

Moreover, the technology enhancement onboard ships rely on the internet between servers, Information Technology (IT) systems and Operation Technology (OT) systems which in turn increase the potential cyber vulnerabilities and risks [3]. OT is hardware and software that monitors and controls processes. Whereas, IT is for information processing, including software, hardware, communications technologies, and related services. In modern ships, the IT and OT systems are vulnerable to attack because they are considered less critical to security and performance [4]. The possible cyberattacks on modern ships, based on already hacked automation systems, were explored [5]. Despite using many cybersecurity measures to protect from threats, it is not adequate because of the continuous growth in technology. It is essential to regularly update the cybersecurity measures aboard ships. The way of attacking ships by cyber attackers changes constantly, and it becomes very difficult to apply cybersecurity measures to prevent them. The

---

V. S. K. Ganesan (✉) · M. Chandra
Indian Maritime University Navi Mumbai Campus, Navi Mumbai, Maharashtra, India
e-mail: veerasenthilkumar@imu.ac.in; mchandra@imu.ac.in

© The Author(s) 2025
S. Bauk (ed.), *Maritime Cybersecurity*, Signals and Communication Technology, https://doi.org/10.1007/978-3-031-87290-7_10

most commonly found cyberattacks on ships include data breaches, ransomware incidents, malware contamination, and invoice fraud. It is mandatory to protect the ship's IT infrastructure from these cyberattacks. It is essential to identify closely the challenges in the implementation of cybersecurity in the shipping industry. The most common challenges are the slow adoption rate, complex systems, ever-changing crew members, inconsistent training, and terrestrial infrastructure dependency. Further, as different types of cyberattacks are continuously evolving in shipping, cyber risk management is mandatory onboard a ship. Cyber risk management is the process of detecting, analyzing, assessing, reporting, and mitigating cyber-related risk to an acceptable level. It helps in formulating new strategies and actions to upgrade security [6]. But it is company and ship specific. However, in 2017, the International Maritime Organization (IMO) adopted resolution MSC.428(98) [7] on maritime cyber risk management in Safety Management System (SMS) and came into force on January 1, 2021. The resolution stated that an approved SMS should consider cyber risk management in accordance with the objectives and functional requirements of the International Safety Management (ISM) code. In addition to the IMO resolution, the U.S. National Institute of Standards and Technology (NIST) Cybersecurity Framework Version 1.1 (April 2018) has also been taken into account in the development of the guidelines. Apart from IMO guidelines, the other guidelines and standards are provided by various associations like International Chamber of Shipping (ICS), International Union of Marine Insurance (IUMI), Baltic and International Maritime Council (BIMCO), Oil Companies International Marine Forum (OCIMF), The International Association of Independent Tanker Owners (INTERTANKO), The International Association of Dry Cargo Shipowners (INTERCARGO), InterManager, Digital Container Shipping Association's (DCSA), and The International Association for Classification Societies (IACS) [8].

By definition, cybersecurity is the body of technologies, processes, and practices designed to protect networks, devices, programs, and data from attack, theft, damage, modification, or unauthorized access. Later, in 1999, Bruce Schneier popularized the concept that cybersecurity is about people, process, and technology [9]. Unlike the process and technology, the human aspect of cybersecurity can mean different things to different people. Traditionally, the human aspect of cybersecurity refers to the risks to an organization, which are created by the people affiliated with the organization, when they interact with technology. But, it refers to both malicious actors and the people who are the cause of the issue. Human behavior is one of the biggest risks to a secure network because understanding typical behavior to identify anomalies is a very challenging issue [10]. Hence, the need for behavioral analysis is increasing in providing cybersecurity solutions and it has become an aspect of cybersecurity education programs [11]. The behavioral analysis deals with understanding the ways of creating risks to the organizations by the individuals and mitigating those risks. While considering the traditional definition of the human aspect of cybersecurity, refers to the risks posed by people, it is important to consider the security-conscious people who implement the additional security to the digital systems. Hence, the human aspect of cybersecurity is no longer just about identifying weaknesses; it is about exploiting the power of our human defenders to enhance the digital security systems to make our organizations more cyber-resilient.

The technical aspect of cybersecurity deals with how to install firewalls and antivirus softwares and encrypt the data. But the main issue of providing cybersecurity arises from the people because the people are the greatest security advocates and weakest security link [12, 13]. Moreover, relying more on intelligent devices encourages cyber criminality such as social engineering, identity theft, and spam emails [14]. Technical attacks differ from social engineering attacks based on the persons involved in the security breach. That is, technical attacks target only the people possessing technical knowledge, but social engineering attacks involve all the employees in an organization [15].

The rest of this chapter deals with the definition of social engineering and its types of attacks, a few cases of social engineering incidents, challenges, and defense strategies of social engineering, cyber issues, and cyber incidents in the maritime sector and finally, the mitigation strategies of maritime cyber incidents.

2 Social Engineering and Types of Attacks

Social engineering is the art of persuading people to reveal confidential information by manipulating their minds rather than technical exploit. The most common forms of social engineering attacks, as depicted in Fig. 1, are phishing, whaling, baiting, diversion theft, business email compromise (BEC), smishing, Quid Pro Quo, pretexting, honeytrap, and tailgating [16, 17].

Phishing is a cyberattack that persuades the users to click a malicious link, download infected files, or reveal personal information through email, phone, SMS, social media, or other forms of personal communication. During COVID-19, everyone spent most of their time online and eventually experienced peak phishing attacks. As per the FBI report, phishing was standing at the top in cybercrimes in 2020, which was approximately double compared to cybercrimes in 2019.

Whaling is a variant of a phishing attack in which the attackers gain access to a user's device or personal information by leveraging personal communication. This type of attack targets mainly the high-level executives in an organization.

Baiting is a type of social engineering attack in which the scammers are giving false promises to deceive the users to reveal personal information or install malware on the system. It could be in the form of online ads or in physical form, like leaving the infected flash drive in an area where the victim could be targeted.

Diversion theft is a cyberattack that started as an offline attack and adapted online to steal confidential information using spoofing the user into sending it to the wrong recipient.

Business Email Compromise (BEC) is a social engineering strategy in which the attackers are impersonating the company's executives and sending emails to their subordinates to execute wire transfers, change banking details, etc. These types of attacks are often difficult to monitor and manage in large organizations.

Fig. 1 Various forms of social engineering attacks

Smishing or SMS-phishing is a SMS-based phishing attack which requires little effort for cyber attackers to execute this attack by purchasing a spoofed number and setting up the malicious link. Today, people spend most of their time on mobile phones and it is easy for attackers to target the victim by sending text messages.

Quid pro quo attack is one in which attackers acquire a user's sensitive information like login credentials of the system or device in exchange for a desirable service. The acquired credentials can then be used by attackers to steal sensitive data stored on the device, and it may be sold on the dark web.

Pretexting is a form of social engineering strategy that involves composing a plausible scenario to convince the victim to divulge sensitive information. It involves impersonating someone in a position of authority and asking questions about the victim to gain their personal information to access their accounts. For example, pretending to be an IT support technician and requesting a user's login credentials over the phone for system maintenance.

Honeytrap is a technique in which the attackers target the individuals, looking for love on online dating websites or social media. Once they identify the victims, they have a relationship with the victim by creating a fake online profile. The attackers take advantage of the relationship and deceive the victim into giving them money, access to personal information, etc.

Tailgating, also known as piggybacking, is a physical breach or unauthorized access by which the attackers gain physical access, using impersonation, to the user's confidential files or devices.

3 Social Engineering Incidents in the Real World

Most of the social engineering attacks occur through technology-based and human-based deception. The former approach is to deceive the user by tricking them to provide confidential information, and the latter approach is to deceive the user by taking advantage of their ignorance [18]. Some of the social engineering attacks that occurred all over the world are discussed in this section [19].

No one can forget the attack on Google and Facebook by Evaldas Rimasauskas, a Lithuanian national, over the period 2013–2015. He deceived the employees of those organizations, by setting up a bogus computer manufacturing company. He trapped them into his trick by sending phishing emails and ransacked more than $100 million. A loss of $75 million was incurred in the attack on Crelan Bank in Belgium in 2016, and it was due to a spear-phishing attack on the CEO.

A voice-phishing attack, called Deepfake attack, was done on the CEO of a UK energy company in March 2019 and ended up in a transfer of $243,000 by the CEO to the scammer. Another massive attack using voice-phishing was done on Twitter employees, in July 2020, in which the attackers hacked the Twitter accounts of many users, including the most famous people. They tweeted to donate to a Bitcoin wallet and ransacked $110,000 in Bitcoin.

Texas Attorney-General warned all the residents in a press release about the popular smishing attack that occurred in September 2020. In this attack, the victim received a fraudulent text message in which he/she was asked to click a link to claim ownership of an undelivered package. Upon clicking the link, the victim was redirected to a malicious website where he/she was asked to disclose his/her personal information.

In April 2021, there was a Business Email Compromise (BEC) scam that targeted the victim to install the malicious code on their device. In this attack, the victim was asked to click an attachment, by which the victim was redirected to a website containing malicious code. The code informs the victim to reenter the login credentials by popping up a false notification. In the same year, a ransomware attack was done on the employees of the UK rail operator, Merseyrail, by an unknown gang. They targeted the employees by sending an email, pretending to be from their Director's email account with the subject "Lockbit Ransomware on Attack and Data Theft."

A data breach, revealing 2096 records of health information and 816 records of personal identification information, occurred in June due to a phishing attack. This attack was accomplished by claiming the login credentials of five employees of Sacramento County in the US. The unfortunate in this attack was that they discovered its occurrence after 5 months. Customers of Oversea-Chinese Bank Corporation (OCBC) in Singapore were attacked by cyber attackers in 2021 and around $8.5 million, across approximately 470 customers, were lost in this scam. The customers of OCBC were trapped by phishing emails to disclose their account details. The situation became worse despite the bank taking efforts to block all the fraudulent domains and alerting the customers. In 2022, phishing emails were sent to the targeted organizations by attackers, pretending to the legitimate emails from the US Department of Labor, to steal Office 365 credentials.

In February 2022, when there was an increasingly unpleasant situation between Russia and Ukraine, Microsoft gave an alert of a new spear-phishing campaign by a Russian hacking team, known as Gamaredon, targeting Ukraine government agencies and NGOs. The starting phase of the attackers depends on sending phishing emails containing malware. The emails sent contain a tracking pixel to confirm the attacker whether has been opened.

A recent incident, Microsoft Azure Outage, occurred on 19 July 2024 due to a faulty update to the Crowdstrike Falcon Sensor software. Following this incident, there are reports of an ongoing phishing campaign targeting Crowdstrike users, leveraging this issue to conduct the following malicious activities, such as sending phishing emails posing as Crowdstrike support to customers, impersonating Crowdstrike staff in phone calls, distributing Trojan malware pretending as recovery tools, etc. [20].

4 Challenges in Preventing Social Engineering Attacks

Technology advancements: As technology grows, cyber attackers use different strategies to deceive ordinary people and high executives. Social engineering attacks are more frequent nowadays and make cybersecurity less effective. Existing techniques to defend against cyberattacks also have fundamental limitations in counteracting them with an ever-growing number of social engineering attacks because they are limited by human subjective decisions. Technology-based techniques fail to provide a complete solution to cybersecurity because of the exploitation of vulnerability in technology.

People innocence: Social engineering is a technique of exploiting the vulnerabilities of humans to accomplish a malicious goal. According to research, around 23% of people accurately manage less than half of the cybersecurity issues and only 4% can handle more than 90% of situations [21].

Growing attacks: Attackers use different types of social engineering attacks like phishing, baiting, whaling, BEC, tailgating, etc. to convince the victim to reveal their or organizations' sensitive information. Devising a cybersecurity mechanism during the design phase to counteract these different types of attacks is a very challenging issue. In today's digital technology era, hackers are stronger than cybersecurity experts. Hence, it is essential for every organization to keep abreast of all their employees about the emerging social engineering techniques to tackle the perils of cyberattacks.

Insider threats: One of the major challenging issues in social engineering is monitoring and handling the people, mostly employees of an organization or the people affiliated to an organization, like contractual or outsourcing workers, who have information regarding assets, security information, and data of an organization. Intentionally or unintentionally, they could exploit the security system of an organization. As a revenge against an organization, for any reason, they cope with attackers toward the decline of the organization's growth.

Budget and Disaster management plan: The knowledge of budget allocation for managing cyberattacks and recovering from them is very important for an organization. Lack of knowledge of the impact of social engineering attacks on organizations is another major challenging issue. Since social engineering attacks occur in different forms, an organization should be prepared to allocate the budget to compensate for any form of attack.

Training: Many organizations focus on technology-based cybersecurity solutions like implementing firewalls, antivirus software, intrusion detection system, etc. They believe that they would provide a holistic solution to cybersecurity problems. They are least bothered about awareness education and training their employees about cybersecurity issues.

5 Defense Strategies of Social Engineering

Social engineers usually target the main assets of organizations, like people, data, software and hardware, and networks. This section deals with the defense mechanisms to prevent social engineering attacks through any of the target points.

The first target point is the people through which the attackers accomplish their goal. Organizations can defend from attacks on people by periodically educating their employees and appointing social engineering technical staff. Information security awareness is essential to reduce the risks associated with information security breaches [22]. Unless employees are guided properly to make the right decisions in the digital world, they will be likely to reveal the organization's data and eventually the loss of business. Hence, the organization should take responsibility for providing information security awareness training to all their employees to keep their cybersecurity systems safe. Every organization must be equipped with adequate social engineering technical staff to predict the possibility of these attacks in all directions and their consequences. They are also required to educate and guide the employees in all situations to stay away safely from these social engineering attacks.

The second target point is the data, which may be at individual level or organization level. Individual level data refers to the personal data of high-profile employees, like their salary, family photos, etc. Organization level data refers to sensitive information like planning documents, financial information, or any private organizational data. The following defense measures are to be adopted by the organization:

Backup and Replication: In order to ensure the integrity and availability of the Confidentiality-Integrity-Availability (CIA) principle, the data of an organization should be periodically backed up and replicated. Employees of the organization must be aware of this back up and replication either on the servers or their working computers.

Principle Of Least Privilege (POLP): This principle limits the level of access to organization data by an employee to execute his task in an organization. The security of the system and data of an organization require correct enforcement of this principle and administrator privileges [23].

Data Sharing Limit: This policy defines the sharing boundaries for an employee inside and outside the organization, i.e., some data can be shared only within a department in the organization, whereas some data can be shared organization level, etc.

Software and hardware are the third target point of the social engineers in an organization. To protect this asset from the social engineering attacks, the organization needs to educate their employees about software and hardware management, email accounts management, authentication policy, and Bring Your Own Device (BYOD) policy. Software and hardware management refers to the software installation, configuration, updates, hardware maintenance plan, etc. Employees must be educated by the IT team of an organization about all these software and hardware management policies. Nowadays, employees of an organization except very few, do not have deeper knowledge of how to maintain their email accounts and do formal way of email communication. Social engineers take advantage of the employee's negligence and send malicious emails by taking control of employees' accounts. Organization must ensure the employees' awareness of how to use and maintain their email accounts by providing well-defined policies. Authentication is one of the important cybersecurity measures to prevent the unauthorized access of one's personal information or organization's sensitive information. It mitigates the problem of impersonation. The authentication process should include more than one form of employees' information like biometric information, complex passwords, captcha, etc. to enhance the security of the organization against social engineering attacks. BYOD policy allows the employees of an organization to bring their own devices like laptops, tablets, etc. for the purpose of official work. However, there are many security risks, which the employees are not aware of, associated with the usage of their own devices. The lack of understanding of BYOD by either the employees or the organizations leads to the loss of critical information resources and assets [24]. Hence, it is essential to ensure the devices are not violating the CIA principle of cybersecurity. The organization should shoulder the responsibility of implementing effective security and privacy policies to manage BYOD.

The network is the fourth target point of the social engineers in an organization. Different networks like LAN, WAN, wired/wireless, etc. are available through which employees of an organization access the databases and servers. All these networks have different security policies that must be known to the employees. Otherwise, it leads to a security threat to the organization. For example, if an employee of an organization attempts to access his organization's local network from his/her friend's computer, without knowing being compromised, then it is a security risk for his/her organization. The knowledge of internet configurations and different network security policies is essential for the employees of an organization to protect the organization's network from social engineering attacks.

6 Cyber Issues in the Maritime Domain

Like other industries, digital transformation in maritime, despite offering substantial benefits, involves more security risks in the cyber domain. IMO adopted Resolution MSC.428(98), 2017 that "encourages administrations to ensure that cyber risks are appropriately addressed in existing safety management systems, as defined in the ISM Code, no later than the first annual verification of the company's Document of Compliance (DOC) after 1 January 2021." Unless cyber awareness is made among all the stakeholders in the shipping industry, risk management cannot be effective. Even though modern technologies can protect against direct attack, the human element is still the most vulnerable in maritime cybersecurity. Performing social engineering attacks on maritime workers than executives is very simple because the executives are well aware of these attacks. The social engineers focus on two groups, namely the crew members who have direct access to the ship's control systems and the shore-based people who have remote access to seaborne networks. On ships, the vulnerable systems for attackers are navigational systems, cargo handling and storage systems, and propulsion and power systems [25]. Among these systems, navigational systems are the most advanced networking and digitally accessible systems onboard. For example, cyber attackers can jam or corrupt the signals from external sensors like GPS, AIS, ARPA, etc., gather critical hydrographic information, and tamper directly with the Electronics Navigational (ENC) chart, if they get access to the Electronic Chart Display and Information System (ECDIS). They can also disable the operating systems of ECDIS stations. Hackers can even control the ship's auto steering algorithm in the advanced integrated bridge navigational systems of modern tanker and passenger ships.

7 Cyber Incidents in the Maritime Sector

In July 2019, the Stena Impero, the UK-flagged oil/chemical tanker, was seized by Iranian forces, as it transited through the Strait of Hormuz. The evidence showed that the Stena Impero received spoofed AIS signals and sent the vessel off course into Iranian waters. This attack resulted in a GPS outage and navigation disruption onboard the vessel [26].

There was a ransomware attack on Compagnie Maritime d'Affrètement Compagnie Générale Maritime (CMA CGM), a French shipping and logistics company, in September 2020 to force the container line to shut down the network and online services [27]. A phishing attack was targeted on a US tugboat via email spoofing in September 2020 [28]. A cyberattack on the Terminal Operating System (TOS) was held in June 2020 to disable completely the port operations from the loading of goods and containers on vessels to the transshipment and entry of cargo to and from the port [29]. A ransomware attack on South Korea's national flagship carrier HMM occurred in 2021 that impacted mainly the email server of the company [30].

The Singapore shipbuilder Sembcorp Marine reported the cyber incident in August 2022. The attack involved an unauthorized party accessing part of its IT network via third-party software products. In this attack, employee data and non-critical information relating to Sembcorp's operations were compromised [31]. Lock Bit cyberattack group attacked the Port of Lisbon on 25 December 2022 and crippled the port operations for 4 days. The attackers claimed the ransom amount of $1.5 million to release the stolen data [32].

In March 2023, the Dutch maritime logistics company, Royal Dirkzwager, confirmed that it was hit with ransomware from the Play Group hackers. This attack did not affect the operations, but data that held a range of contracts and personal information from the servers was compromised [33]. The operation of the German-based shipyard, the Lürssen shipyard, was paralyzed by the BianLian APT group in April 2023, and data was stolen. The background is supposed to be a phishing attack that served as a precursor to a ransomware attack [34]. The LockBit 3.0 ransomware group crippled the container operations across all terminals within the Nagoya Port, one of the largest ports in Japan, in July 2023. The investigation confirmed that the operator received a ransom demand in exchange for the recovery of the port's system [35]. In November 2023, an attack on DP World Australia, an international firm specializing in cargo logistics and port operations, was reported. The attack forced the port operator to close the terminals in Sydney, Melbourne, Brisbane, and Fremantle [36].

The above-said cyber incidents with a detailed descriptions are depicted in Table 1.

Analysis of the above cyber incidents, reported from 2019 to 2023, shows that most of the incidents occurred due to phishing, malware, and ransomware. The reason behind these attacks was the vulnerability of cybersecurity awareness. It is observed that inadequate knowledge and awareness of cybersecurity among the crew members and officers onboard the ship are major causes of all these attacks.

Table 1 Maritime cyber incidents with detailed description

Time period of the incident	Victim	Targeted system (IT/OT)	Nature of the attack involved	Description
July 2019	Stena Impero, the UK flagged oil/chemical tanker	OT	GPS Spoofing	Received spoofed AIS signals and sent the vessel off course into Iranian waters. Resulted in a GPS outage and navigation disruption onboard the vessel
September 2020	CMA CGM, the French shipping and logistics company	IT	Phishing	Targeted on US tug boat via email spoofing and forced the container line to shut down the network and online services
June 2020	Iran's TOS at Shahid Rajaee Seaport	IT, OT	Malware attack	Disabled completely the port operations from loading of goods and containers on vessels to the transshipment and entry of cargo to and from the port. Also, infiltrated and damaged a number of private operating systems at the ports
June 2021	HMM, the South Korea's national flagship carrier	IT	Ransomware	Impacted mainly the email server of the company
August 2022	Sembcorp Marine, the Singapore shipbuilder	IT	Hacking	Stolen employee data and non-critical information relating to Sembcorp's operations
December 2022	The Port of Lisbon	IT	Ransomware	Claimed a ransom amount of $1.5 million to release the stolen data
March 2023	Royal Dirkzwager, the Dutch maritime logistics company	IT	Ransomware	Not affected the operations, but a range of contracts and personal information from the servers was compromised
April 2023	Lürssen shipyard, the German-based shipyard	IT	Phishing	Paralyzed the operation of the shipyard and the data was stolen
July 2023	The Nagoya Port, Japan	IT	Ransomware	Crippled the container operations across all terminals within the Nagoya Port by the LockBit 3.0 ransomware group
November 2023	DP World Australia, the international logistics management services	IT	Seemed to be Ransomware	No ransom demand from the attacker, but forced to close the terminals in Sydney, Melbourne, Brisbane, and Fremantle

8 Mitigation Strategies of Maritime Cyber Incidents

Like other sectors, the maritime sector also suffers from techno-based cyberattacks like GPS spoofing, AIS attacks, GNSS attacks, supply chain attacks, DDoS attacks, etc., malware/ransomware attacks, and social engineering attacks. The two main reasons behind all these attacks are the lack of adequate security mechanisms onboard the vessel [5] or in the port and the lack of knowledge of cybersecurity by the employees/crew members onboard. The first reason could be compensated by strengthening the framework in IT/OT systems onboard the vessel or in the port through installing strong antivirus software and firewall system, implementing intrusion detection and prevention system, segmenting the network in case of any attack, periodic backing up of data to either external storage device or cloud server and other data protection and recovery strategies. Despite having strong IT policies on board, the second reason is very challenging to compensate, because it relies on how much effective training is provided by the owner of the vessel and how much is perceived by the employees/crew members. Some of the strategies to prevent social engineering attacks are as follows [37]:

- Deploying Multi-factor Authentication (MFA) to avoid identity theft.
- Auto-scanning of any external device, in the case of inserting it into a digital system onboard.
- Enabling spam filters to protect inboxes.
- Auto-updating all the application software as well as the system software to patch the vulnerabilities.
- Penetration testing to exploit vulnerabilities.
- Identifying the possible critical assets other than product/service/intellectual property.
- Implementing risk-based WAF to avoid any potential infiltration.
- Ensuring the secured communication through SSL certificates.
- Accessing only secured website which start with *https://*.

Apart from the above-said strategies, some of the best practices are to be followed by the crew members and the officers onboard a vessel [38] as well as the employees in port operations, to reduce the problem of cyber risks.

- Use of strong passwords to protect the computers/digital systems available onboard and their personal accounts.
- Update all the application/system software regularly.
- Back up the sensitive data regularly.
- Understand the Cyber Netiquettes to know how to communicate with others in cyberspace.
- Know how to react in case of any cyber risk.
- Not to use any external device, without scanning it, and public Wi-Fi.
- Not to click on any suspicious links and offers from unknown people on the internet.
- Not to store sensitive information in an unprotected drive.

9 Conclusion

A holistic review of cyberattacks across the globe and especially in the maritime domain, either in the port operations or onboard a vessel, is presented in this chapter to understand the necessity of enhancing cybersecurity to meet the requirements of the maritime industry, as technology is proliferating at a rapid pace. The exploitation of human vulnerabilities by cyber attackers, through social engineering via phishing, baiting, whaling, malware, etc., is found to be more common than the technical vulnerabilities exploitation in the current scenario because of inadequate knowledge of the crew members in handling cyber issues and insufficient IT infrastructure facilities. This shows that the maritime industry is at risk and it may worsen the situation, as the future of the maritime involves new technologies like autonomous vessels, remotely operated vessels, port digitalization, etc., unless a proper cybersecurity mechanism has to be adopted to ensure cyber resilience onboard as well as off-ship. Based on the review of cyber incidents and challenges in the maritime industry, it is concluded that ship owners, higher authorities in ports, OT/IT architects, ship builders; classification societies (maritime), regulatory bodies, and insurance companies should shoulder the responsibility to provide guidelines and awareness about the cyber risks and threats to the IT infrastructure deployed onboard the ship as well as the off-shore to the crew members and employees. It is encouraged to provide them with effective training about the various ways of cyberattacks in different situations, by having a partnership with third-party cybersecurity solution providers, in order to protect them from the tricks of attackers.

Acknowledgments Research for this publication was funded by the EU Horizon2020 project 952360-MariCybERA.

References

1. K.D. Jones, K. Tam, M. Papadaki, Threats and impacts in maritime cyber security. Eng. Technol. Ref. **1**(1) (2016). https://doi.org/10.1049/etr.2015.0123
2. DNV, Maritime cyber security (n.d.), https://www.dnv.com/maritime/insights/topics/maritime-cyber-security/. Accessed 27 Jun 2024
3. P.H. Meland, K. Bernsmed, E. Wille, Ø.J. Rødseth, D.A. Nesheim, A retrospective analysis of maritime cyber security incidents. TransNav **15**(3), 519–530 (2021)
4. Y. Gu, J.C. Goez, M. Guajardo, S.W. Wallace, Autonomous vessels: state of the art and potential opportunities in logistics. Int. Trans. Oper. Res. **28**, 1706–1739 (2021)
5. F. Akpan, G. Bendiab, S. Shiaeles, S. Karamperidis, M. Michaloliakos, Cybersecurity challenges in the maritime sector. Network **2**, 123–138 (2022). https://doi.org/10.3390/network2010009
6. Lotus Containers, Best practices for cybersecurity in the maritime industry (2023), https://www.lotus-containers.com/en/best-practices-for-cybersecurity-in-the-maritime-industry/. Accessed 27 Jun 2024

7. International Chamber of Shipping, The guidelines on cyber security onboard ship (n.d.), https://www.ics-shipping.org/wp-content/uploads/2020/08/guidelines-on-cyber-security-onboard-ships-min.pdf. Accessed 1 Jul 2024
8. IMO, Maritime cyber risk (n.d.), https://www.imo.org/en/OurWork/Security/Pages/Cyber-security.aspx. Accessed 1 Jul 2024
9. J. Barker, *The Human Nature of Cybersecurity* (Educause, 2019) https://er.educause.edu/articles/2019/5/the-human-nature-of-cybersecurity
10. C. Asan, The role of cyber situational awareness of humans in social engineering cyber attacks on the maritime domain. Mersin Univ. J. Marit. Faculty **5**(2), 22–36 (2023)
11. M. Rowley, *How Human Behavior Affects Cybersecurity* (Columbia Southern University, 2021), https://www.columbiasouthern.edu/blog/blog-articles/2021/february/human-aspects-of-cyber-security/
12. Inmarsat Maritime Security Services. Cyber security requirements for IMO 2021 [White paper] (2020), https://www.inmarsat.com/content/dam/inmarsat/corporate/documents/maritime/insights/Cyber_Security_IMO2021_Requirements_WhitePaper_3.2.pdf
13. G.L. Orgill, G.W. Romney, M.G. Bailey, P.M. Orgill, The urgency for effective user privacy-education to counter social engineering attacks on secure computer systems, in *CITC5 '04: Proceedings of the 5th Conference on Information Technology Education, UT, Salt Lake City, USA*, (2004)
14. M.A.B. Farah, E. Ukwandu, H. Hindy, D. Brosset, M. Bures, I. Andonovic, X. Bellekens, Cyber-security in the maritime industry: a systematic survey of recent advances and future trends. Information **13**(22) (2022). https://doi.org/10.3390/info13010022
15. H. Aldawood, G. Skinner, Educating and raising awareness on cyber security social engineering: a literature review, in *2018 IEEE International Conference on Teaching, Assessment, and Learning for Engineering, Wollongong, NSW, Australia*, (2018)
16. R.M. Abdulla, A.H. Faraj, C.O. Abdullah, A.H. Amin, T.A. Rashid, Analysis of social engineering awareness among students and lecturers. IEEE Access **11**, 101098 (2023). https://doi.org/10.1109/ACCESS.2023.3311708
17. B.L. Bergmans, 10 Types of social engineering attacks and how to prevent them (2023), https://www.crowdstrike.com/cybersecurity-101/types-of-social-engineering-attacks/
18. I.A.M. Abass, Social engineering threat and defense: a literature survey. J. Inf. Secur. **9**, 257–264 (2018). https://doi.org/10.4236/jis.2018.94018
19. Tessian, 15 examples of real social engineering attacks (2023), https://www.tessian.com/blog/examples-of-social-engineering-attacks/
20. Certin, Phishing Campaign leveraging CrowdStrike outage event (2024), https://cert-in.org.in/s2cMainServlet?pageid=PUBADV01&CACODE=CICA-2024-3308
21. C. Beaman, A. Barkworth, T.D. Akande, S. Hakak, M.K. Khan, Ransomware: recent advances, analysis, challenges and future research directions. Comput. Secur. **111**, 102490 (2021). https://doi.org/10.1016/j.cose.2021.102490
22. N.A.G. Arachchilage, S. Love, Security awareness of computer users: a phishing threat avoidance perspective. Comput. Hum. Behav. **38**, 304–312 (2014). https://doi.org/10.1016/j.chb.2014.05.046
23. R. Heartfield, G. Loukas, A taxonomy of attacks and a survey of defence mechanisms for semantic social engineering attacks. ACM Comput. Surv. (CSUR) **48**(3), 1–39 (2015)
24. A.B. Garba, J. Armarego, D. Murray, W. Kenworthy, Review of the information security and privacy challenges in bring your own device (BYOD) environments. J. Inf. Privacy Secur. **11**(1), 38–54 (2015)
25. L. Vashchenko, Perils of a new dimension: socially engineered attacks in maritime cyber-security (2021), https://cimsec.org/perils-of-a-new-dimension-socially-engineered-attacks-in-maritime-cybersecurity/
26. M.W. Bockmann, Seized UK tanker likely 'spoofed' by Iran (2019), https://lloydslist.com/LL1128820/Seized-UK-tanker-likely-spoofed-byIran#:~:text=THE%20UK%2Dflagged%20product%20tanker,transited%20the%20Strait%20of%20Hormuz

27. C. Shen, J. Baker, CMA CGM confirms ransomware attack (2020), https://lloydslist.com/LL1134044/CMA-CGM-confirms-ransomware-attack
28. Dryad Global, Tug owners warned after first detected cyber attack (2020), https://channel16.dryadglobal.com/tug-owners-warned-after-first-detected-cyber-attack. Accessed 13 Jul 2024
29. Hellenic Shipping News, How Iran's Shahid Rajaee seaport was cyber-attacked (2020), https://www.hellenicshippingnews.com/how-irans-shahid-rajaee-seaport-was-cyber-attacked/. Accessed 13 Jul 2024
30. Heimdal, South Korean Company HMM reveals it had suffered a cyberattack on its Email servers (2021), https://heimdalsecurity.com/blog/hmm-reveals-it-had-suffered-a-cyberattack-on-its-email-servers/. Accessed 13 Jul 2024
31. M. Wingrove, Sembcorp marine addresses cyber-security incident (2022), https://www.rivieramm.com/news-content-hub/news-content-hub/sembcorp-marine-addresses-cyber-security-incident-72723. Accessed 13 Jul 2024
32. C. Tsoukas, Maritime cybersecurity attacks on the rise (n.d.), https://marpoint.gr/blog/maritime-cybersecurity-attacks-on-the-rise/. Accessed 13 Jul 2024
33. J. Greig, Dutch shipping giant Royal Dirkzwager confirms Play ransomware attack (2023), https://therecord.media/royal-dirkzwager-ransomware-attack-dutch-shipping
34. B2B Cyber Security, Cyber attacks on North German shipyards (2023), https://b2b-cyber-security.de/en/cyberangriffe-auf-norddeutsche-werften/. Accessed 13 Jul 2024
35. E. Hayden, Maritime security: 2023's attacks, pitfalls and impacts (n.d.), https://pacmar.com/article/maritime-security-2023s-attacks-pitfalls-and-impacts/. Accessed 13 Jul 2024
36. D.H. Kass, Cyberattack launched against DP World Australia Port (2023), https://www.msspalert.com/news/does-dp-world-australia-hack-signal-uptick-in-critical-infrastructure-attacks
37. C. Vinugayathri, 10 Ways businesses can prevent social engineering attacks (2020), https://www.indusface.com/blog/10-ways-businesses-can-prevent-social-engineering-attacks/
38. T. Coquil, G. Poupard, Best practices for cyber security on-board ships (n.d.), https://cyber.gouv.fr/sites/default/files/2017/06/best-practices-for-cyber-security-on-board-ships_anssi.pdf. Accessed 13 Aug 2024

Open Access This chapter is licensed under the terms of the Creative Commons Attribution-NonCommercial-NoDerivatives 4.0 International License (http://creativecommons.org/licenses/by-nc-nd/4.0/), which permits any noncommercial use, sharing, distribution and reproduction in any medium or format, as long as you give appropriate credit to the original author(s) and the source, provide a link to the Creative Commons license and indicate if you modified the licensed material. You do not have permission under this license to share adapted material derived from this chapter or parts of it.

The images or other third party material in this chapter are included in the chapter's Creative Commons license, unless indicated otherwise in a credit line to the material. If material is not included in the chapter's Creative Commons license and your intended use is not permitted by statutory regulation or exceeds the permitted use, you will need to obtain permission directly from the copyright holder.

Toward Secure Marine Navigation: A Deep Learning Framework for Radar Network Attack Detection

Md. Alamgir Hossain, Md. Delwar Hossain, Latifur Khan, Hideya Ochiai, Md. Saiful Islam, and Youki Kadobayashi

1 Introduction

Marine radar systems are critical for safe navigation and collision avoidance in maritime environments. However, these systems are increasingly vulnerable to a variety of sophisticated cyberattacks that can disrupt their functionality and compromise maritime safety [14]. The primary attacks targeting marine radar systems include DoS attacks, which can freeze or disable radar operations; transformation attacks, such as scaling, rotation, and translation, which distort radar images; and object manipulation attacks, which involve adding, removing, or relocating objects on the radar display [6]. These attacks can create false navigational information, leading to potentially catastrophic consequences.

The security of marine radar systems is imperative to ensure safe and efficient maritime navigation. Addressing the vulnerabilities in these systems through rigorous research offers multiple benefits. Firstly, enhancing radar security will significantly reduce the risk of maritime accidents caused by cyberattacks, thereby

M. A. Hossain · M. S. Islam
Institute of Information and Communication Technology (IICT), Bangladesh University of Engineering and Technology (BUET), Dhaka, Bangladesh

M. D. Hossain (✉)
Angelo State University, San Angelo, TX, USA
e-mail: delwar.hossain@angelo.edu

L. Khan
The University of Texas at Dallas, Richardson, TX, USA

H. Ochiai
The University of Tokyo, Tokyo, Japan

Y. Kadobayashi
Division of Information Science, Nara Institute of Science and Technology, Ikoma, Nara, Japan

protecting human lives and the environment [11]. Secondly, it will safeguard the economic interests of the shipping industry by preventing disruptions and ensuring the continuity of maritime trade routes [19]. Thirdly, it will bolster national security by protecting critical maritime infrastructure from potential threats [12]. Researching and developing advanced detection and mitigation strategies for radar cyberattacks will not only improve the robustness of maritime operations but also contribute to the broader field of cybersecurity, setting new standards for protecting critical systems [1, 8].

Solving the problem of securing marine radar systems against cyberattacks is inherently challenging due to several factors. Firstly, the complexity and diversity of attack vectors make it difficult to develop a one-size-fits-all solution. Attacks can range from DoS to sophisticated transformation and object manipulation attacks, each requiring different detection and mitigation strategies. Secondly, the maritime environment is highly dynamic and involves continuous movement and varying conditions, which complicates the identification of anomalous behavior. Thirdly, the data generated by radar systems is vast and multidimensional, necessitating advanced techniques to process and analyze it in real time. Additionally, there is a lack of comprehensive datasets and simulation environments that accurately represent real-world scenarios, hindering the development and testing of effective security solutions. Lastly, implementing robust security measures must not interfere with the operational efficiency of radar systems, posing a further challenge in balancing security and performance. These complexities demand innovative approaches and extensive research to develop effective and practical solutions [5, 7, 9].

To address the complex challenge of securing marine radar systems against cyberattacks, this research developed a comprehensive approach that leverages advanced deep learning techniques. We leverage the RadarPWN dataset [20], utilizing packet capture files to extract comprehensive features with proper preprocessing. We then train a 1D Convolutional Neural Network (1D CNN) specifically designed for detecting and classifying network-based attacks on marine radar systems. This approach surpasses previous methods by offering both binary and multi-class classification, enabling precise identification of specific attack types. Furthermore, the integration of early stopping and hyperparameter optimization ensures high accuracy and robustness. Unlike traditional methods, our deep-learning-based solution can adapt to evolving threats and process vast amounts of data in real time, providing a scalable and efficient defense mechanism that sets a new benchmark in maritime cybersecurity. The key contributions of this research are listed below:

– At first introduced a deep-learning-based framework for detecting and classifying network attacks on marine radar systems, setting a new standard for maritime cybersecurity.
– Explored a 1D Convolutional Neural Network (1D CNN) tailored for both binary and multi-class classification of radar cyber-threats, enhancing the accuracy and reliability of maritime cybersecurity measures.

- Achieved stable and high accuracy in detecting various types of radar cyber-threats.
- Set a new benchmark in maritime cybersecurity, showcasing the efficacy of deep learning in protecting critical maritime infrastructure.

In the next sections of this chapter, we present the literature review, followed by the methodology, results and discussion, conclusion, and references.

2 Related Works

In this section, the related research is reviewed with a focus on their methodologies, strengths, limitations, objectives, data sources, key findings, novelty, and relevance to provide a comprehensive understanding of the current landscape in maritime cybersecurity.

Longo et al. [13] investigated cyberattacks on maritime radar systems, proposing a model that integrates electronic countermeasures (ECM) with cyber-false flags to simulate ECM effects. They used the ASTERIX CAT-240 protocol in a realistic maritime scenario to assess the feasibility of these attacks. Key findings showed how cyber-false flags could mislead attribution and complicate detection efforts. The research's strengths include a comprehensive threat model and realistic simulations, while limitations involve potential real-world variability. This study's novelty lies in combining cyber-false flags with ECM, highlighting the need for advanced maritime cybersecurity measures. In 2024 Giacomo et al. [15] introduce a threat model and attack methodology that leverage domain-specific vulnerabilities in maritime Supervisory Control and Data Acquisition (SCADA) systems [21], specifically targeting the steering gear system on ships. They develop a three-phase attack methodology—reconnaissance, weaponization, and delivery—using process mining techniques to reverse engineer operational procedures and craft targeted malware capable of executing precise attacks on shipboard ICS. However, the controlled environment in which the experiments were conducted may not fully represent real-world conditions. Additionally, the research assumes a significant level of attacker expertise and access to onboard systems, which may not always be realistic.

Amro et al. [3] present a systematic approach for analyzing anomalies in navigational NMEA [4] messages, identifying possible malicious causes, and proposing appropriate detection algorithms. They develop a tool, NMEA-Manipulator, to test their methodology through various cyberattack scenarios on sensor data. They evaluate the proposed approach using two use cases: traditional Integrated Navigation System (INS) and Autonomous Passenger Ship (APS) [2, 16]. They employ specification and frequency-based detection methods to identify anomalies with high confidence. But it relies on simulated environments, which may not fully capture real-world complexities and sensor variances. Additionally, the approach focuses on NMEA0183 messages, potentially overlooking threats targeting other standards like NMEA2000.

A machine learning method for classifying maritime targets and detecting anomalies using Doppler signatures from millimeter wave radars introduced by Rahman et al. in 2023 [18]. The authors present a comprehensive dataset collected from field trials and demonstrate the application of supervised and unsupervised learning techniques for effective target classification. The research utilizes Doppler spectrum and spectrogram features from 77, 94, and 207 GHz radars to train supervised learning models (including support vector machines) for multi-class and binary classification, achieving validation accuracies up to 93.3% and 95%, respectively. An unsupervised one-class SVM for anomaly detection is also employed. It focuses on low sea states and low grazing angles, limiting the generalizability of the results to other maritime environments. Additionally, the computational load of the methodology may challenge real-time implementation on small- to medium-sized vessels, as it relies heavily on high-dimensional feature extraction and processing. The classification algorithms used, particularly support vector machines, may struggle with high-dimensional data and require significant tuning to achieve optimal performance, with accuracy potentially decreasing in more complex, real-world scenarios.

Existing approaches for marine radar security, such as traditional approaches and anomaly detection techniques, are limited by their inability to fully adapt to real-world variability and the rapid evolution of cyber-threats. As technology is updating day by day, the ways attackers target marine radar systems change rapidly. These methods often require extensive tuning and struggle with high-dimensional data, resulting in potential security gaps. Our proposed deep learning approach, using a 1D CNN, overcomes these limitations by efficiently handling complex patterns and continuously learning from new data. This adaptability ensures robust defense against emerging threats, providing superior accuracy and resilience compared to traditional methods.

3 Proposed Approach

In this section, the methodology of the research is described, encompassing the processing of .pcap files for feature extraction, data preprocessing, model development, classification process, and the evaluation metrics used. The pipeline of the proposed approach is shown in Fig. 1, and the model development with the classification process using advanced 1D CNN is illustrated in Fig. 2.

3.1 Dataset Preparation

In this research, we utilize the RadarPWN dataset [20], which consists of meticulously recorded packet capture files simulating a variety of attack and benign scenarios in marine radar systems.

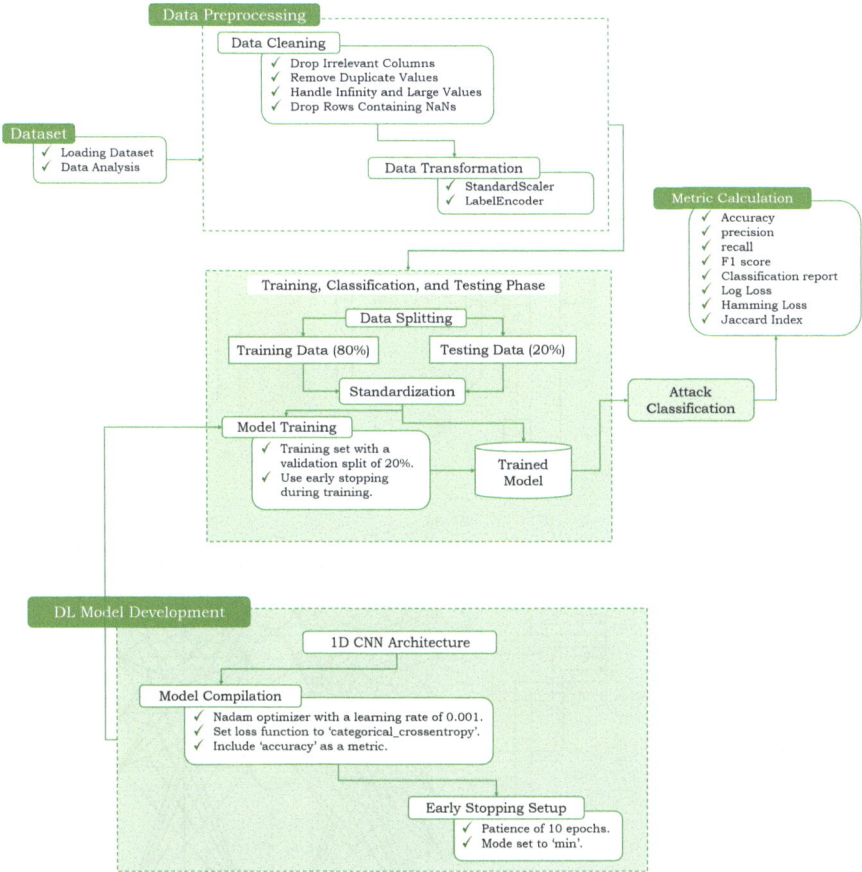

Fig. 1 Pipeline of the proposed approach

The RadarPWN dataset was developed to simulate and capture network-based cyberattacks against marine radar systems within modern Integrated Bridge Systems (IBS). This dataset was recorded using a sophisticated simulation environment that includes the BridgeCommand ship simulator, the Radar Attack Tool (RAT), and the OpenCPN chart plotter. The dataset comprises network traffic from Navico BR24 and NMEA 0183, recorded under both benign and attack scenarios.

The dataset includes 72 captures, each representing a unique scenario generated and recorded three times with slight variations in the ship's route. The scenarios cover three simulated environments: RadarRostock, RadarSantaCatalina, and RadarSimpleEstuary. The attack classes in these scenarios include benign, Denial of Service (DoS), scaling, rotation, translation, addition, removal, and relocation. Each attack class is represented by various specific attack techniques such as random pixels, image rotation, and ship addition/removal.

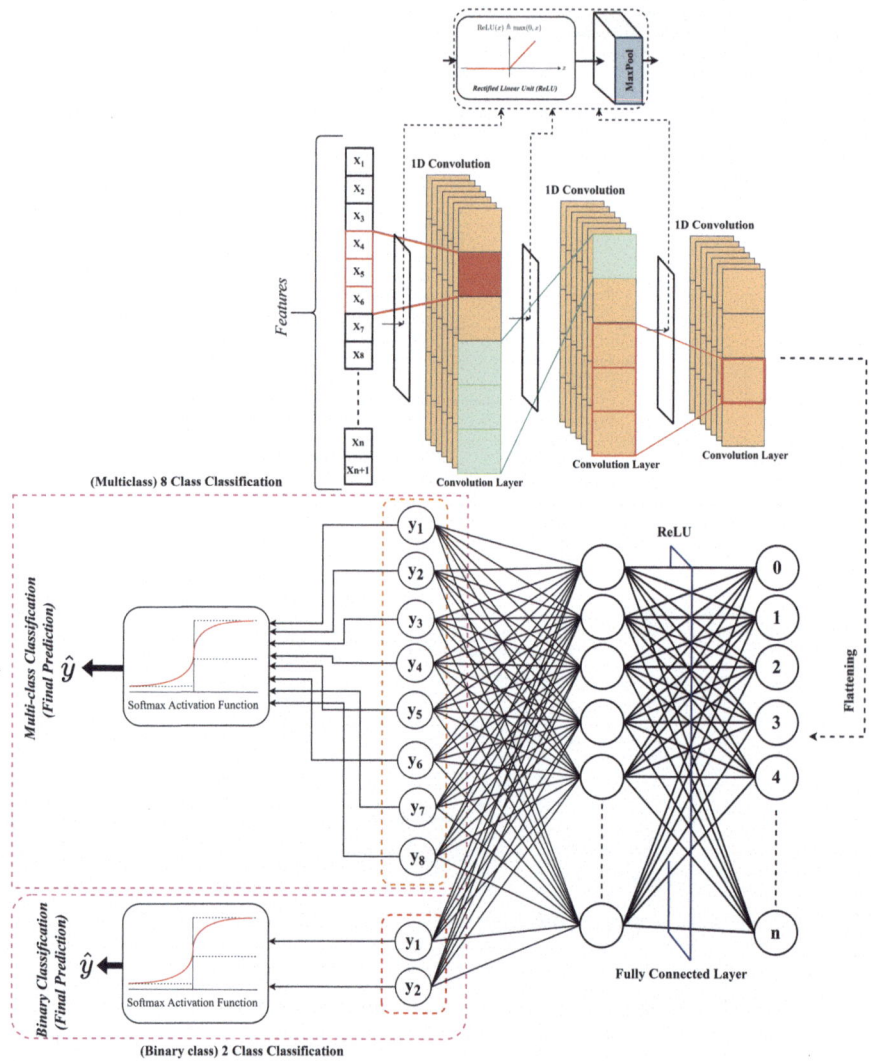

Fig. 2 Classification process of 1D CNN

The packet captures in the dataset record the communication between OpenCPN and RAT, with UDP packets carrying NMEA 0183 sentences and Navico BR24 radar data. Key tools used include the BridgeCommand simulator and the OpenCPN chart plotter, with protocols such as NMEA 0183 and Navico BR24 being integral to the network traffic recorded.

The feature extraction process from .pcap files begins by reading the packet capture files using the `rdpcap` function, which loads the raw network traffic data. Initially, an empty list is created to store individual packet data, and a dictionary

Toward Secure Marine Navigation 201

is initialized to maintain flow statistics. Each packet is examined to determine if it contains the IP layer, and relevant information such as source and destination IPs, ports, protocol, timestamp, and packet length is extracted. The flow ID, which uniquely identifies each communication flow based on source and destination details, is computed to aggregate packets belonging to the same flow. This step also involves distinguishing between forward and backward packets based on the IP addresses.

Once the packets are categorized, flow statistics are updated by tracking the start and end times of each flow, calculating flow duration, and accumulating packet counts and lengths for both directions. For each packet, several packet-level features are computed, including inter-arrival times, header lengths, and flag counts. These features help in capturing the behavior of the traffic flow. Subsequently, additional flow-level statistics such as packet length mean, variance, active and idle times, and down/up ratio are calculated to provide a comprehensive view of the traffic characteristics.

The data is then labeled by initializing a "Label" column to "Benign" and applying predefined labels based on specific scenarios, marking periods when attacks occur. The final steps involve converting the list of packet data into a DataFrame, cleaning the data by removing rows with NaN values, and dropping columns with constant values across all rows. The processed DataFrame is then saved to a CSV file for further analysis. This structured and detailed feature extraction process transforms raw network traffic data into a format suitable for machine learning models, enabling effective detection and classification of cyberattacks on marine radar systems. The feature extraction process resulted in a dataset containing 1,843,991 rows and 66 columns, including the "Label" column, which was stored in the CSV file.

3.2 Data Preprocessing

Data preprocessing is a crucial step in the machine learning pipeline, as it ensures the quality and consistency of the data used for model training and evaluation. Proper preprocessing transforms raw data into a clean and structured format, enabling the model to learn effectively and make accurate predictions. This step includes handling missing values, normalizing features, and encoding categorical variables, which collectively enhance the performance and robustness of the model [10].

After loading the .csv file, any infinity values and large numerical representations are replaced with NaNs, followed by the removal of all rows containing NaN values by Eqs. 1–3. This ensures the dataset is clean and free from outliers.

$$df = df.replace([+\infty, -\infty], NaN) \tag{1}$$

$$df = df.replace(regex\_pattern, NaN) \tag{2}$$

$$df.dropna(inplace = True) \quad (3)$$

where X represents the dataset, ∞ and NaN are infinity and not-a-number values, respectively, and `regex_pattern` is a regular expression pattern for large numerical representations.

Then, the numerical and categorical columns are separated. Numerical columns are normalized using the StandardScaler, which transforms the data to have a mean of 0 and a standard deviation of 1 with Eq. 4.

$$X_{norm} = \frac{X - \mu}{\sigma} \quad (4)$$

where X_{norm} is the normalized data, X is the original data, μ is the mean, and σ is the standard deviation.

Next, categorical columns are ensured to be in string format and encoded using LabelEncoder. Label encoding assigns each unique category in a column to an integer value.

$$X_{cat} = \text{Encode}(X_{cat}) \quad (5)$$

where X_{cat} is the categorical data and `Encode` represents the label encoding process.

After, the data is sampled with 10% from each class for balanced training. The features (X) and labels (y) are separated, and the labels are encoded and converted into categorical format for multi-class classification. The one-hot encoding converts class labels into a binary matrix using the following Eq. 6:

$$y_{\text{one-hot}} = \text{OneHotEncode}(y) \quad (6)$$

where X_{cat} is the categorical data and `Encode` represents the label encoding process. In the next step, the dataset is split into training and testing sets, with the training set comprising 80% and the testing set comprising 20% of the data. The numerical features are standardized to ensure the model training is not skewed by differing scales of the features. Then the training data is scaled with Eq. 7.

$$X_{\text{train\_scaled}} = \frac{X_{\text{train}} - \mu_{\text{train}}}{\sigma_{\text{train}}} \quad (7)$$

where $X_{\text{train\_scaled}}$ is the scaled training data, X_{train} is the original training data, μ_{train} is the mean of the training data, and σ_{train} is the standard deviation of the training data.

$$X_{\text{test\_scaled}} = \frac{X_{\text{test}} - \mu_{\text{train}}}{\sigma_{\text{train}}} \quad (8)$$

where $X_{\text{test\_scaled}}$ is the scaled testing data, X_{test} is the original testing data, μ_{train} is the mean of the training data, and σ_{train} is the standard deviation of the training data.

The standardized data is then reshaped for input into the 1D CNN model, ensuring the data is in the correct format for model training:

$$X_{\text{train\_cnn}} = \text{Reshape}(X_{\text{train\_scaled}}, N_{\text{train}}, F, 1) \qquad (9)$$

$$X_{\text{test\_cnn}} = \text{Reshape}(X_{\text{test\_scaled}}, N_{\text{test}}, F, 1) \qquad (10)$$

where $X_{\text{train\_cnn}}$ and $X_{\text{test\_cnn}}$ are the reshaped training and testing data, respectively, N_{train} and N_{test} are the number of training and testing samples, respectively, and F is the number of features.

3.3 Deep Learning Model Development

1D Convolutional Neural Networks (1D CNNs) are a type of deep learning model specifically designed to process and analyze sequential data [17]. Unlike traditional neural networks, CNNs can capture spatial hierarchies by applying convolutional filters to the input data, making them highly effective for tasks such as time series analysis, natural language processing, and in this case, network traffic classification. The 1D CNN architecture typically includes layers such as convolutional layers, pooling layers, flattening layers, and dense layers, each contributing to feature extraction, dimensionality reduction, and classification. In our proposed approach, a 1D CNN is utilized to classify network traffic data into different types of attacks and benign traffic. This method leverages the hierarchical feature extraction capabilities of CNNs to improve the accuracy and robustness of network security models. The architectural values of 1D CNN are shown in Table 1. The parameters used with their values are displayed in Table 2, and the classification process is illustrated in Fig. 2.

Table 1 Architecture of the 1D convolutional neural network model. Values include layer type, output shape, and the number of parameters

Layer (type)	Output shape	Param #
conv1d (Conv1D)	(None, 63, 64)	256
max_pooling1d (MaxPooling1D)	(None, 31, 64)	0
conv1d_1 (Conv1D)	(None, 29, 128)	24704
max_pooling1d_1 (MaxPooling1D)	(None, 14, 128)	0
conv1d_2 (Conv1D)	(None, 12, 256)	98560
max_pooling1d_2 (MaxPooling1D)	(None, 6, 256)	0
flatten (Flatten)	(None, 1536)	0
dense (Dense)	(None, 128)	196736
dense_1 (Dense)	(None, 8) for Multi, (None, 2) for Binary	1032

Table 2 Used parameter with the value of developed 1D CNN

Parameter	Value
Conv1D layers	3
Kernel size	3
Activation function (Conv1D and Dense)	**ReLU**
Pooling size	2
Output Activation Function	**Softmax**
Optimizer	**Nadam**
Learning rate	0.001
Loss function	Categorical cross-entropy
Metrics	Accuracy
Early stopping monitor	Validation loss
Patience for early stopping	10
Batch size	32
Number of epochs	200
Validation split	0.2

The classification process of the proposed 1D CNN for classifying different attacks is described below:

3.3.1 Convolutional Layers

The convolutional layers apply filters to the input data to extract local patterns. Each filter w slides over the input sequence x, performing a convolution operation:

$$(x * w)[t] = \sum_{i=1}^{K} x[t + i - 1]w[i] \tag{11}$$

where K is the kernel size, t is the position in the input sequence, and w represents the filter weights. The activation function applied is the ReLU (Rectified Linear Unit):

$$\text{ReLU}(z) = \max(0, z) \tag{12}$$

Three convolutional layers are used with filter sizes of 64, 128, and 256, respectively, each followed by a ReLU activation function.

3.3.2 Pooling Layers

After each convolutional layer, a MaxPooling layer is applied to downsample the feature maps, reducing their dimensionality and retaining the most significant features. The pooling operation is defined as

$$\text{MaxPooling}(x)[t] = \max_{i=1}^{P} x[t+i-1] \tag{13}$$

where P is the pooling size.

3.3.3 Flattening Layer

The output from the last pooling layer is flattened into a single vector:

$$\text{Flatten}(x) = x \tag{14}$$

This layer prepares the data for the fully connected dense layers.

3.3.4 Dense Layers

The dense layers perform the final classification by combining the features extracted by the convolutional and pooling layers. The first dense layer applies a linear transformation followed by a ReLU activation function:

$$\text{Dense}(x) = \text{ReLU}(Wx + b) \tag{15}$$

where W is the weight matrix, x is the input vector, and b is the bias vector. The output layer uses a softmax activation function to produce the probabilities for each class:

$$\text{Softmax}(z)_i = \frac{e^{z_i}}{\sum_{j=1}^{C} e^{z_j}} \tag{16}$$

where C is the number of classes, and z is the input to the softmax function.

3.3.5 Training and Optimization

The model is compiled using the Nadam optimizer and categorical cross-entropy loss function. The Nadam optimizer is a variant of the Adam optimizer that incorporates Nesterov momentum:

$$m_t = \beta_1 m_{t-1} + (1 - \beta_1) g_t \tag{17}$$

$$v_t = \beta_2 v_{t-1} + (1 - \beta_2) g_t^2 \tag{18}$$

$$\hat{m}_t = \frac{m_t}{1 - \beta_1^t} \tag{19}$$

$$\hat{v}_t = \frac{v_t}{1 - \beta_2^t} \tag{20}$$

$$\theta_t = \theta_{t-1} - \eta \frac{\hat{m}_t}{\sqrt{\hat{v}_t} + \epsilon} \tag{21}$$

where m_t and v_t are the first and second moment estimates, β_1 and β_2 are the decay rates, g_t is the gradient, η is the learning rate, and ϵ is a small constant to prevent division by zero.

3.3.6 Early Stopping

Early stopping is employed to prevent overfitting. The training process monitors the validation loss and stops when it does not improve for a specified number of epochs (patience), restoring the best model weights.

3.4 Evaluation Metrics

Evaluation metrics are crucial in assessing the performance and effectiveness of a deep learning model. They provide quantitative measures to understand how well the model predicts outcomes, identifies different classes, and distinguishes between benign and attack traffic. The evaluation metrics used in this research are described in Table 3.

The proposed 1D CNN model effectively extracts features from network traffic data and classifies different types of attacks with high accuracy by leveraging convolutional layers, pooling layers, and dense layers, optimized with the Nadam optimizer and monitored with early stopping.

4 Results and Discussion

In this section, the results of the research are described with proper figures and tables to illustrate the effectiveness of the proposed approach in detecting and classifying cyberattacks on marine radar systems.

Table 3 Metrics for model evaluation

Metric name	Description	Equation
Accuracy	The ratio of correctly predicted instances to the total instances. High accuracy indicates the model's overall effectiveness in classifying both benign and attack traffic	$\frac{TP+TN}{TP+TN+FP+FN}$
Precision	The ratio of correctly predicted positive observations to the total predicted positives. High precision indicates fewer false positives, crucial for minimizing false alarms in attack detection	$\frac{TP}{TP+FP}$
Recall	The ratio of correctly predicted positive observations to all observations in the actual class. High recall indicates fewer false negatives, ensuring that most attacks are detected	$\frac{TP}{TP+FN}$
F1-Score	The harmonic mean of precision and recall. High F1-Score balances precision and recall, providing a single metric for model performance, especially when dealing with imbalanced datasets	$2 \times \frac{\text{Precision} \times \text{Recall}}{\text{Precision} + \text{Recall}}$
ROC AUC	The area under the Receiver Operating Characteristic curve, plotting the true positive rate against the false positive rate. High AUC indicates the model's ability to distinguish between classes, with 1.0 representing perfect classification	$\int_0^1 \text{TPR}(f)\,df$

4.1 Experimental Setup

The entire implementation of our proposed approach for marine radar security was conducted using Google Colaboratory (Google Colab), a cloud-based platform that offers a flexible and efficient environment for developing and executing machine learning models. We used Python 3 as the programming language, along with key libraries such as Pandas for data manipulation, NumPy for numerical computations, Matplotlib for plotting, Scikit-learn for data preprocessing and model evaluation, and TensorFlow/Keras for building and training the 1D Convolutional Neural Network (1D CNN) model. The RadarPWN dataset, including .pcap files for raw network traffic data, was converted into .csv files for feature extraction.

4.2 Results for Both Binary and Multi-classification

Figure 3 illustrates the distribution of different classes within the dataset used for training and evaluating our 1D CNN model for detecting and classifying cyberattacks on marine radar systems. The pie chart shows the proportion of each class, with the *Benign* class constituting 22.3% of the dataset, while the remaining classes—*DoS, Scaling, Addition, Rotation, Move, Translation,* and *Removal*—each make up between 8.8 and 11.9% of the dataset. For the training of the 1D CNN

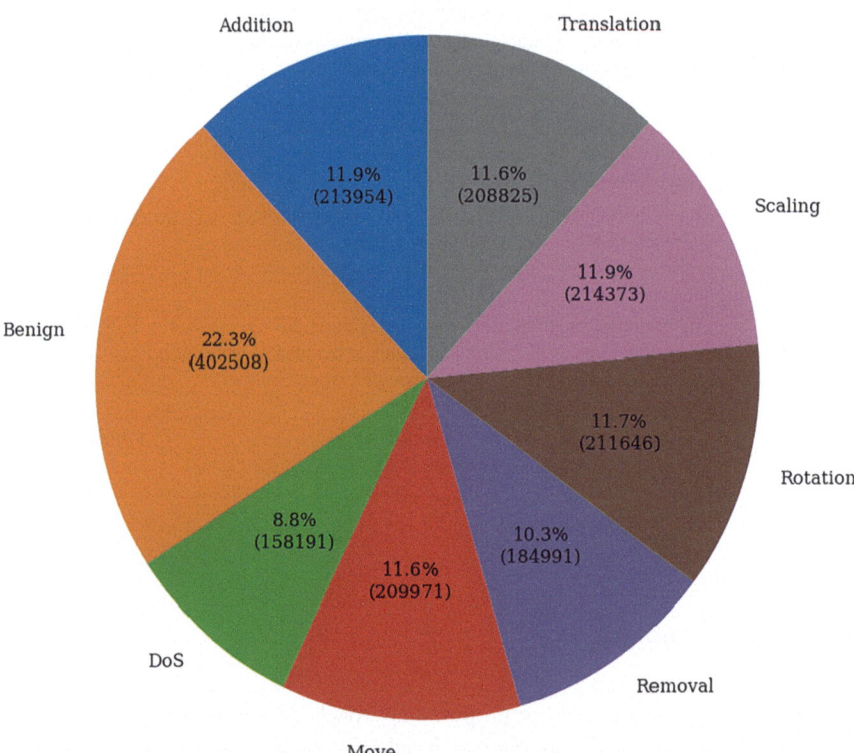

Fig. 3 Amount of extracted data for each class

model in multi-class and binary classification, 10% of the data from each class was used. For binary classification, all attack types were consolidated into a single *Attack* class, allowing the model to distinguish between benign and malicious traffic effectively. This balanced and comprehensive dataset ensures that the model can accurately detect and classify various types of cyber-threats to marine radar systems.

Table 4 presents evaluation metrics for binary classification tasks, distinguishing between different "Attack Type" and "Benign" data. The proposed 1D CNN model for binary classification in marine radar security exhibits outstanding performance, with perfect evaluation metrics for the majority of attack types and nearly perfect metrics for DoS attacks. This indicates the model is highly effective in correctly identifying both benign and attack instances across various types of attacks, showcasing its robustness and reliability in a practical security context.

Figure 4 shows the confusion matrix for the binary classification task, where the 1D CNN model distinguishes between *Attack*(Translation, Removal, DoS) and *Benign* classes. The matrix demonstrates the model's performance in correctly

Table 4 Evaluation metrics for binary (attack type and benign) classification

Class type	Accuracy	Precision	Recall	F1-score
Scaling	1.0000	1.0000	1.0000	1.0000
Addition	1.0000	1.0000	1.0000	1.0000
Rotation	1.0000	1.0000	1.0000	1.0000
Move	1.0000	1.0000	1.0000	1.0000
Translation	1.0000	1.0000	1.0000	1.0000
Removal	1.0000	1.0000	1.0000	1.0000
DoS	0.9935	0.9936	0.9935	0.9935

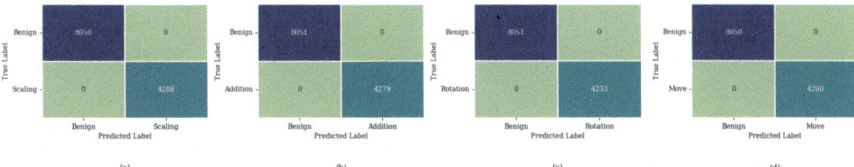

Fig. 4 Confusion matrix for binary class (Translation, Removal, DoS)

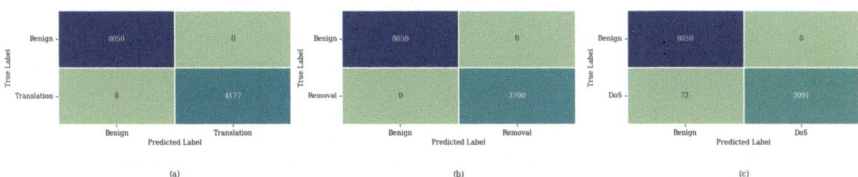

Fig. 5 Confusion matrix for binary class (Scaling, Addition, Rotation, Move)

identifying and classifying the test data. Notably, the model achieves perfect classification with no false positives or false negatives, as indicated by the zero values in the off-diagonal entries. This result underscores the model's exceptional accuracy and effectiveness in detecting and differentiating between benign traffic and various cyberattacks on marine radar systems.

Figure 5 consists of three confusion matrices representing the performance of the binary classification model for three types of attacks: Translation, Removal, and DoS. The confusion matrices demonstrate that the proposed 1D CNN model performs exceptionally well in classifying between benign instances and different types of attacks, with perfect classification for Translation and Removal attacks. For these attack types, there are no misclassifications, indicating 100% accuracy, precision, recall, F1 score, and Jaccard Index. However, for DoS attacks, while the model still performs excellently, it shows a slight decrease in performance with 73 instances of false positives, leading to a slightly lower accuracy.

Table 5 provides the classification report for a multi-class classification model. The multi-class classification model demonstrates outstanding performance, achieving perfect scores (1.0000) in Precision, Recall, and F1 Score across all class types except for a marginally lower recall (0.9900) for the DoS class. With an overall accuracy of 1.00 and impeccable macro and weighted averages, the model

Table 5 Classification report for multi-class classification

Class type	Precision	Recall	F1 score
Addition	1.0000	1.0000	1.0000
DoS	1.0000	0.9900	1.0000
Move	1.0000	1.0000	1.0000
Removal	1.0000	1.0000	1.0000
Rotation	1.0000	1.0000	1.0000
Scaling	1.0000	1.0000	1.0000
Translation	1.0000	1.0000	1.0000
Accuracy	1.0000		
Macro Avg	1.0000	1.0000	1.0000
Weighted Avg	1.0000	1.0000	1.0000

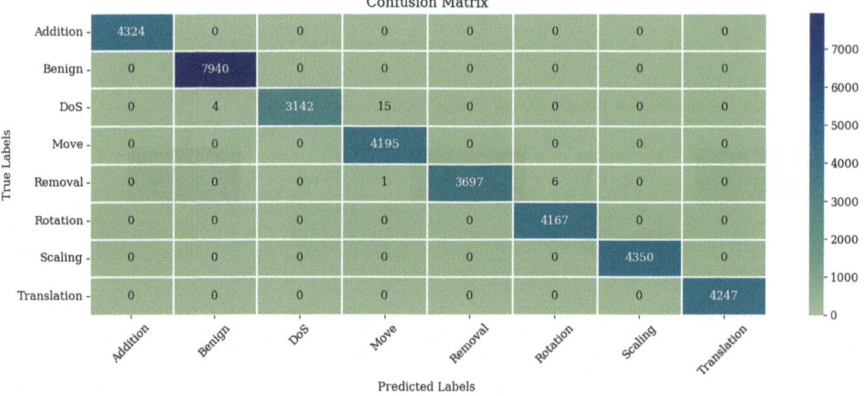

Fig. 6 Confusion matrix for multi-class

showcases exceptional robustness and reliability, positioning it as a leading solution in the realm of marine radar security.

Figure 6 presents the confusion matrix for the multi-class classification task, where the 1D CNN model distinguishes between various attack types on marine radar systems and benign traffic. The model demonstrates high accuracy, correctly classifying all instances of *Addition*, *Benign*, *Move*, *Rotation*, *Scaling*, and *Translation*. For *DoS* attacks, 3,142 instances are correctly classified, with minimal misclassifications (4 as *Benign* and 15 as *Move*). For *Removal* attacks, 3,697 instances are correctly classified, with only 1 misclassified as *Benign* and 6 as *Rotation*. The few misclassifications underscore the model's effectiveness and robustness in accurately detecting and classifying multiple types of cyberattacks on marine radar systems.

Figure 7 shows the training and validation loss and accuracy curves for the 1D CNN model during the binary classification task. The loss curves indicate a rapid decrease to near-zero values, demonstrating effective learning. The accuracy curves quickly rise to around 100% and remain stable, confirming the model's excellent performance in distinguishing between benign and attack traffic. The close

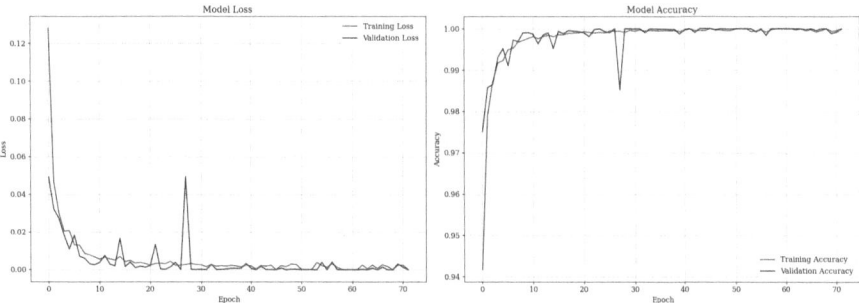

Fig. 7 Model loss and accuracy for binary classification (addition)

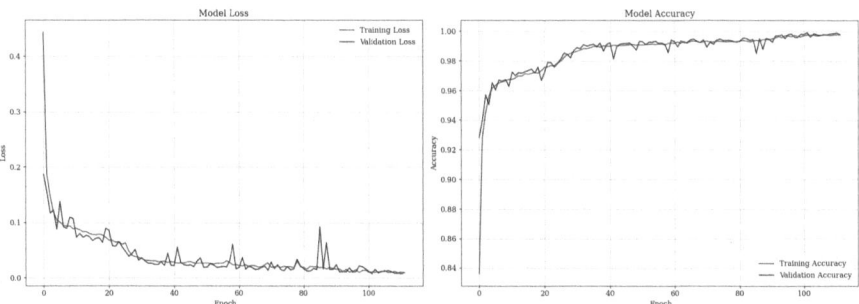

Fig. 8 Model loss and accuracy for multi-class classification

alignment of training and validation metrics underscores the model's robustness and reliability in real-world scenarios, affirming the efficacy of our approach in detecting cyberattacks on marine radar systems.

Figure 8 shows the training and validation loss and accuracy curves for the 1D CNN model during the multi-class classification task. The loss curves indicate a rapid decline and stabilization at low values, with minimal fluctuations, demonstrating effective learning and good generalization. The accuracy curves quickly rise and stabilize near 100%, with close alignment between training and validation metrics, indicating high model performance. These results confirm the model's robustness and reliability in accurately classifying various types of cyberattacks on marine radar systems.

Figure 9 shows the Receiver Operating Characteristic (ROC) curve for the binary classification task, illustrating the model's ability to distinguish between benign and attack traffic. The ROC curve demonstrates a perfect classification performance, with an area under the curve (AUC) of 1.0000. The curve rises sharply to the top-left corner, indicating a high true positive rate and a low false positive rate, underscoring the model's exceptional accuracy and reliability in detecting cyberattacks on marine radar systems.

Figure 10 illustrates the Receiver Operating Characteristic (ROC) curves for the multi-class classification task, showing the model's performance in distinguishing

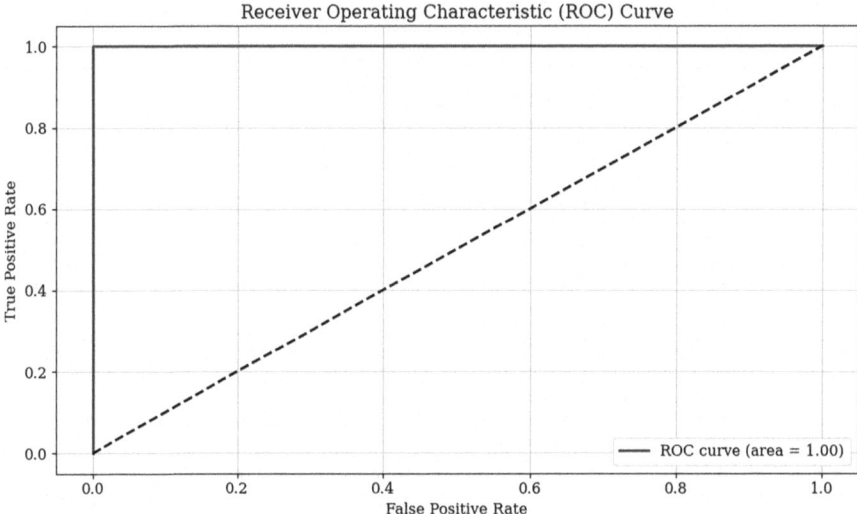

Fig. 9 ROC curve of the model for binary classification (addition)

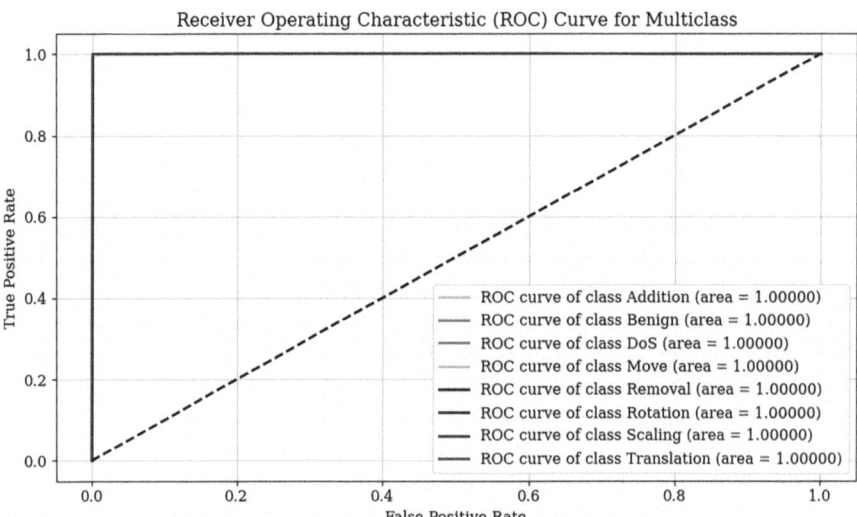

Fig. 10 ROC curve of the model for multi-class classification

between different attack types and benign traffic. Each class's ROC curve reaches the top-left corner, indicating perfect classification with an area under the curve (AUC) of 1.0000 for all classes. This exceptional performance demonstrates the model's ability to accurately identify and differentiate between multiple types of cyberattacks on marine radar systems.

The proposed 1D CNN approach for marine radar security shows strong performance across both binary and multi-class classification tasks, highlighting its potential effectiveness in this domain. This improved accuracy can be attributed to the model's ability to efficiently extract and process complex patterns from high-dimensional network traffic data. However, the approach relies on simulated datasets, which may not fully capture the complexities and variabilities of real-world maritime environments.

5 Conclusion

In this research, we explored the security of marine radar systems by developing a robust 1D Convolutional Neural Network (1D CNN) model to detect various network-based cyberattacks. We meticulously processed the RadarPWN dataset, which includes packet capture files of both benign and attack scenarios, and applied comprehensive preprocessing techniques to ensure data quality. The proposed 1D CNN effectively extracted hierarchical features from the network traffic data, demonstrating high accuracy and reliability in classifying different types of attacks. The significance of the results lies in the model's exceptional performance, achieving nearly perfect scores across multiple evaluation metrics, thereby enhancing the security of marine radar systems. This research not only highlights the potential of deep learning in cybersecurity but also sets a foundation for future research. Future directions include extending this approach to other maritime systems, exploring more sophisticated attack vectors, and incorporating real-time detection capabilities to further bolster marine cybersecurity defenses.

Acknowledgment Research for this chapter was also supported by the EU Horizon 2020 project MariCybERA (Agreement No. 952360).

References

1. F. Akpan, G. Bendiab, S. Shiaeles, S. Karamperidis, M. Michaloliakos, Cybersecurity challenges in the maritime sector. Network **2**(1), 123–138 (2022)
2. A. Amro, V. Gkioulos, S, Katsikas, Communication architecture for autonomous passenger ship. Proc. Institut. Mech. Eng. Part O J. Risk Reliab. **237**(2), 459–484 (2021)

3. A. Amro, A. Oruc, V. Gkioulos, S. Katsikas, Navigation data anomaly analysis and detection. Information **13**(3), 104 (2022)
4. C. Boudehenn, O. Jacq, M. Lannuzel, J.-C. Cexus, A. Boudraa, Navigation anomaly detection: An added value for maritime cyber situational awareness, in *2021 International Conference on Cyber Situational Awareness, Data Analytics and Assessment (CyberSA)* (2021), pp. 1–4
5. S. Cohen, T. Gluck, Y. Elovici, A. Shabtai, Security analysis of radar systems, in *CPS-SPC'19: Proceedings of the ACM Workshop on Cyber-Physical Systems Security & Privacy* (2019), pp. 3–14
6. S. Cohen, E. Levy, A. Shaked, T. Cohen, Y. Elovici, A. Shabtai, RadArnomaly: protecting radar systems from data manipulation attacks. Sensors **22**(11), 4259 (2022)
7. M. Erbas, S.M. Khalil, L. Tsiopoulos, Systematic literature review of threat modeling and risk assessment in ship cybersecurity. Ocean Eng. **306**, 118059 (2024)
8. M. Farah, E. Ukwandu, H. Hindy, D. Brosset, M. Bures, I. Andonovic, X. Bellekens, Cyber security in the maritime industry: a systematic survey of recent advances and future trends. Information **13**, 22 (2022)
9. A.J. Fenton, Preventing catastrophic cyber–physical attacks on the global maritime transportation system: a case study of hybrid maritime security in the Straits of Malacca and Singapore. J. Marine Sci. Eng. **12**(3), 510 (2024)
10. M.A. Hossain, M.S. Islam, Ensuring network security with a robust intrusion detection system using ensemble-based machine learning. Array **19**, 100306 (2023)
11. K. Jones, K. Tam, K., M. Papadaki, Threats and Impacts in Maritime Cyber Security. Eng. Technol. Ref. https://doi.org/10.1049/etr.2015.0123
12. T. Liebetrau, C. Bueger, Advancing coordination in critical maritime infrastructure protection: lessons from maritime piracy and cybersecurity. Int. J. Crit. Infrastruct. Protect. **46**, 100683 (2024)
13. G. Longo, A. Merlo, A. Armando, E. Russo, Electronic attacks as a cyber false flag against maritime radars systems, in *2023 IEEE 48th Conference on Local Computer Networks (LCN)* (2023), p. 1–6
14. G. Longo, E. Russo, A. Armando, A. Merlo, Attacking (and defending) the maritime radar system. IEEE Trans. Inf. Forens. Secur. **PP**, 1–1 (2023)
15. G. Longo, F. Lupia, A. Pugliese, E. Russo, Physics-aware targeted attacks against maritime industrial control systems. J. Inf. Secur. Appl. **82**, 103724 (2024)
16. Z. Pietrzykowski, P. Wołejsza, Ł. Nozdrzykowski, P. Borkowski, P. Banaś, J. Magaj, J. Chomski, M. Maka, S. Mielniczuk, A. Pańka, P. Hatłas-Sowińska, E. Kulbiej, M. Nozdrzykowska, The autonomous navigation system of a sea-going vessel. Ocean Eng. **261**, 112104 (2022)
17. E.U.H. Qazi, A. Almorjan, T. Zia, A one-dimensional convolutional neural network (1d-CNN) based deep learning system for network intrusion detection. Appl. Sci. **12**(16), 7986 (2022)
18. S. Rahman, A. Vattulainen, D. Robertson, Machine learning-based approach for maritime target classification and anomaly detection using millimetre wave radar Doppler signatures. IET Radar Sonar Navigat. **18**, 344–360 (2023)
19. G.M.S. Rashid, M. Ikram, Safeguarding maritime trade corridors: the crucial role of infrastructure development and port security measures. Int. J. Contemporary Issues Soc. Sci. **3**(2), 2727–2736 (2024)
20. K. Wolsing, A. Saillard, J. Bauer, E. Wagner, C. Sloun, I. Fink, M. Schmidt, K. Wehrle, M. Henze, Network attacks against marine radar systems: A taxonomy, simulation environment, and dataset, in *2022 IEEE 47th Conference on Local Computer Networks (LCN)* (IEEE, Piscataway, 2022), pp. 114–122
21. M. Zaghloul, Online ship control system using supervisory control and data acquisition (SCADA). Int. J. Comput. Sci. Appl. **3**, 6 (2014)

Open Access This chapter is licensed under the terms of the Creative Commons Attribution-NonCommercial-NoDerivatives 4.0 International License (http://creativecommons.org/licenses/by-nc-nd/4.0/), which permits any noncommercial use, sharing, distribution and reproduction in any medium or format, as long as you give appropriate credit to the original author(s) and the source, provide a link to the Creative Commons license and indicate if you modified the licensed material. You do not have permission under this license to share adapted material derived from this chapter or parts of it.

The images or other third party material in this chapter are included in the chapter's Creative Commons license, unless indicated otherwise in a credit line to the material. If material is not included in the chapter's Creative Commons license and your intended use is not permitted by statutory regulation or exceeds the permitted use, you will need to obtain permission directly from the copyright holder.

Improving Security and Privacy with Raspberry Pi Devices

Radoje Džankić, Zvonko Bulatović, and Sanja Bauk

1 Introduction

Raspberry Pi, a powerful and versatile single-board computer, offers a number of hacking capabilities to improve security and privacy [1–5]. Using its capabilities, users can create robust firewalls, encrypt their data, and effectively monitor network traffic. Raspberry Pi offers several options that can be adapted to individual needs. Here are some key considerations and the best options available. Creating a VPN (Virtual Private Network), for instance, allows users to establish a secure and encrypted connection between their device and the Internet. This ensures that all data transmitted over the network remains private and protected from potential eavesdroppers. The Raspberry Pi can be used as a VPN server, providing a reliable and cost-effective solution for improving privacy. By setting up a VPN on the Raspberry Pi, users can securely access their home network remotely or browse the Internet anonymously. While setting up a home surveillance system, a Raspberry Pi, with its small size and low power consumption, is an ideal platform for building a home surveillance system. By connecting the camera module to the Raspberry Pi, users can remotely monitor their premises and improve security. Various software options are available, such as MotionEyeOS and ZoneMinder, which provide comprehensive monitoring functions and can be easily installed on the Raspberry Pi. The Raspberry Pi can also be used as a secure file server to store and share sensitive data within a local network. By configuring the Raspberry Pi as a NAS (Network Attached Storage) device, users can create their own cloud storage solution with

---

R. Džankić (✉) · S. Bauk
Estonian Maritime Academy, Tallinn University of Technology, Tallinn, Estonia
e-mail: radoje.dzankic@taltech.ee; sanja.bauk@taltech.ee

Z. Bulatović
Ministry of Internal Affairs, Podgorica, Montenegro
e-mail: zvonko.bulatovic@gov.me

© The Author(s) 2025
S. Bauk (ed.), *Maritime Cybersecurity*, Signals and Communication Technology, https://doi.org/10.1007/978-3-031-87290-7_12

full control over their data. Options like OpenMediaVault and Nextcloud offer users robust security values, ensuring data privacy and accessibility. It's worth noting that while Raspberry Pi hacks can significantly improve security and privacy, it is essential to adopt best practices to maximize their effectiveness.

2 Data Protection

Here are given some recommendations for the Raspberry Pi's data protection.

Regularly updating the Raspberry Pi operating system and software. Keeping the system up-to-date ensures that all security vulnerabilities are patched, reducing the risk of exploitation by malicious actors.

Enabling Two-Factor Authentication (2FA). Implementing 2FA adds an extra layer of security by requiring users to provide an additional verification step, such as a code sent to their mobile device, when accessing their Raspberry Pi or connected services.

Using strong passwords. Using unique and complex passwords for Raspberry Pi accounts and services is key to preventing unauthorized access. Password managers can help generate and securely store strong passwords.

It is crucial to comprehend the value of security and privacy in the technologically advanced world of today. Raspberry Pi hacks provide a range of ways to enhance these features, from creating a home security system to configuring a VPN for safe online browsing. People can take charge of their digital security and privacy by utilizing the Raspberry Pi's capabilities and best practices, which will reduce risks and provide security.

3 Securing the Raspberry Pi

When it comes to securing your Raspberry Pi, there are some essential tips and best practices that can help protect your device and your data. Whether you are a novice or an experienced user, it is crucial that you take the necessary precautions to ensure the safety of your Raspberry Pi. In this section, we will explore some key steps you can take to improve the security of your Raspberry Pi.

Change the default password. One of the first and most critical steps to secure your Raspberry Pi is to change the default password. By default, the Raspberry Pi comes with a generic password, which makes it vulnerable to unauthorized access. Changing your password is as simple as running the "passwd" command in the terminal and following the prompts. Remember to choose a strong and unique password that includes a combination of letters, numbers, and special characters.

Keep your Raspberry Pi up-to-date. Regularly updating the software of your Raspberry Pi is essential to maintain security. Updates often include security patches that address vulnerabilities and bugs. To update your Raspberry Pi, open a terminal and run "sudo apt update" followed by "sudo apt upgrade." This ensures that your device is updated and protected from potential threats.

Enable firewall protection. Enabling a firewall on your Raspberry Pi can add an extra layer of security by filtering incoming and outgoing network traffic. The default firewall utility for the Raspberry Pi is "iptables." To enable the firewall, open a terminal and run "sudo apt install iptables-persistent." Follow the instructions to configure the firewall rules according to your specific needs. It is recommended to block all unnecessary incoming connections and allow only the necessary ones.

Disable SSH password authentication. By default, SSH (Secure Shell) allows password-based authentication, which can be vulnerable to brute-force attacks. To improve security, it is recommended to disable SSH password authentication and use SSH keys instead. Generate an SSH key pair on your local machine and copy the public key to your Raspberry Pi. Then disable password authentication by editing the SSH configuration file using "sudo nano/etc/ssh/sshd_config" and changing "PasswordAuthentication" to "no." Restart the SSH service for the changes to take effect.

Enable two-factor authentication (2FA). Implementing two-factor authentication adds an extra layer of security to your Raspberry Pi. Users are required to provide two forms of identification, typically a password and a temporary code from a mobile app or hardware token. To enable 2FA, you can use apps like Google Authenticator or Authy. Install the required packages and follow the instructions to set up 2FA for your Raspberry Pi login.

Use a VPN for remote access. If you need to remotely access your Raspberry Pi, it is crucial that you do so securely. Using VPN can encrypt your connection and provide a secure tunnel between your device and the Raspberry Pi. There are various VPN options available, such as OpenVPN or WireGuard. Choose a VPN protocol and set it up on your Raspberry Pi by following the instructions provided.

Back up your data regularly. Backing up your data is essential to ensure you can recover in the event of any security incident or hardware failure. Consider making regular backups of your Raspberry Pi SD card or important files to an external storage device or cloud storage service. Tools like "rsync" or "dd" can be used to automate the backup process and ensure the integrity of your data.

Securing the Raspberry Pi is a key step in protecting your device and data from potential threats. By following these basic tips and best practices, you can significantly improve the security of your Raspberry Pi and enjoy a worry-free experience. Security is an ongoing process, so be vigilant and keep up with the latest security updates and practices to ensure the long-term security of your Raspberry Pi.

4 Setting Up VPN on Raspberry Pi: Anonymizing the Internet Connection

With the increasing prevalence of cyber threats and constant surveillance of online activities, it is imperative to take proactive measures to protect our Internet connection. One effective way to achieve this is to set up a VPN on the Raspberry Pi. Raspberry Pi is a small, affordable, and versatile computer that can be used for a variety of purposes, including improving security and privacy. By configuring a VPN on your Raspberry Pi, you can encrypt your Internet traffic and route it through a secure server, effectively anonymizing your online activities.

Understanding VPNs. Before getting into the process of setting up a VPN on your Raspberry Pi, it is important to understand what a VPN is and how it works. A VPN creates a secure and encrypted connection between your device and the Internet. It acts as an intermediary, directing your Internet traffic through a remote server and masking your IP address. This ensures that your online activities are protected from prying eyes, whether it's your Internet service provider, hackers, or government surveillance.

Choosing the right VPN protocol. When setting up a VPN on your Raspberry Pi, one of the first decisions you'll need to make is choosing the appropriate VPN protocol. There are several options available, each with their own strengths and weaknesses. Some of the most common VPN protocols include OpenVPN, WireGuard, and PPTP. OpenVPN is widely regarded as the most secure and reliable option, offering excellent encryption and compatibility. WireGuard, on the other hand, is a relatively new protocol known for its simplicity and speed. PPTP, although fast, is considered less secure and should be avoided if possible.

Installation and configuration of the VPN server. After choosing a VPN protocol that suits your needs, the next step is to install and configure the VPN server on your Raspberry Pi. Various software options are available, such as PiVPN, OpenVPN Access Server, and Algo VPN. The PiVPN is a popular choice due to its simplicity and user interface. It simplifies the installation process and provides an intuitive web-based management interface. OpenVPN Access Server offers more advanced functions and is ideal for larger scale deployments. Algo VPN, on the other hand, provides a self-hosted VPN solution that prioritizes security and privacy.

Setting up VPN clients. After setting up the VPN server on your Raspberry Pi, you will need to configure the VPN clients on your devices to connect to the server. Fortunately, most VPN protocols are supported by a wide range of operating systems, including Windows, macOS, Linux, iOS, and Android. This allows you to secure all your devices, ensuring your Internet connection remains private and anonymous, regardless of the platform you use. Simply install the appropriate VPN client software, enter your server information, and connect to start enjoying the benefits of VPN.

Additional considerations. While setting up a VPN on your Raspberry Pi can greatly improve your security and privacy, there are a few additional considerations to keep in mind. First, it is important to choose a reliable VPN service provider if you opt for a third-party VPN server. Look for vendors that prioritize privacy, have a strong reputation, and offer robust encryption. In addition, regularly updating your Raspberry Pi operating system and VPN software is essential to ensure you benefit from the latest security patches and improvements. Finally, consider the performance effect of using a VPN, as the encryption and routing of your Internet traffic may result in slightly slower speeds.

Setting up a VPN on the Raspberry Pi is an effective way to anonymize the Internet connection and improve security and privacy. By understanding the different VPN protocols available, choosing the right software, and properly configuring servers and clients, one can enjoy a secure and private online experience. Remember to choose a reliable VPN service provider, keep your software up-to-date, and be aware of the potential impact on performance. With these considerations in mind, we can get the most out of the Raspberry Pi and take control over our own digital privacy.

5 Protecting the Network

In an interconnected world, where our homes, offices, and even our cars are connected to the Internet, the need for a robust and reliable network security system has become paramount. With the rise of cyber threats and the increasing vulnerability of our personal data, it is imperative to take proactive measures to protect our networks from intruders. One such measure is building a Raspberry Pi firewall, a cost-effective and versatile solution that can significantly improve the security and privacy of your network.

Raspberry Pi firewall. The Raspberry Pi firewall is essentially a network security device that monitors and controls incoming and outgoing network traffic based on predefined security rules. It acts as a barrier between your network and the outside world, filtering potentially harmful traffic and preventing unauthorized access to your network. The Raspberry Pi, with its low power consumption and compact size, makes an ideal platform for building a firewall that can be easily customized and adapted to your specific security requirements.

Choosing the right firewall software. When it comes to choosing firewall software for your Raspberry Pi, there are several options to be considered. Some popular choices include:

- iptables: This is a built-in firewall tool in Linux that provides a robust and flexible solution for filtering network traffic. It offers granular control over network packets and can be configured to create complex firewall rules.

However, this requires a deep understanding of networking concepts and command-line configuration.
- UFW (Uncomplicated Firewall): As the name suggests, UFW is a user-friendly frontend for iptables that simplifies the configuration process. It provides an easy-to-use command-line interface and offers preconfigured profiles for common services, making it an excellent choice for beginners.
- pfSense: If one prefers a more advanced feature-rich firewall solution, pfSense is worth considering. It is a free and open-source firewall distribution based on FreeBSD and offers a web-based graphical interface for easy configuration. It provides advanced functions such as VPN support, traffic shaping, and intrusion detection, making it suitable for more complex network settings.

After considering the pros and cons of each option, iptables stands out as the best choice for building a Raspberry Pi firewall. Although it requires some technical expertise, it offers unparalleled flexibility and control over network traffic filtering. Setting up a firewall. After choosing firewall software, the next step is to set up the Raspberry Pi as a dedicated firewall device. However, building a Raspberry Pi firewall is not a one-time task. Namely, it requires constant monitoring and maintenance to ensure its effectiveness.

6 Protecting Files and Communications

Raspberry Pi, a small but powerful computer, can change the game when it comes to encrypting data and improving security. Here, we will explore the various methods and tools available for encrypting data with the Raspberry Pi.

Using encryption software. One of the easiest ways to encrypt data on a Raspberry Pi is to use encryption software. There are various options available, each with their own strengths and weaknesses. Here are some choices:

- VeraCrypt: This open-source disk encryption software allows you to create encrypted containers or encrypt entire partitions. It offers strong encryption algorithms and supports multiple platforms, making it a versatile choice.
- GnuPG: Also known as GPG, this command-line tool provides cryptographic privacy and authentication for data communication. With GPG, you can encrypt files, email, and even create digital signatures. Although it requires little technical knowledge, GPG offers robust encryption capabilities.
- Cryptomator: If you are looking for a user-friendly option, Cryptomator is worth considering. This open-source software creates a virtual disk drive where you can store your files, encrypting them on the fly. It integrates seamlessly with cloud storage services, ensuring your data stays safe even in the cloud.

The VPN setup. The VPNs are effective means of securing the communications and protecting privacy. By encrypting your Internet traffic and routing it through a remote server, VPNs protect your data from prying eyes. Raspberry Pi can be transformed into a VPN server, allowing you to establish a secure connection wherever you are. Here are some options to be considered:

- OpenVPN: This open-source VPN protocol is widely appreciated for its strong security features. With OpenVPN, you can create a secure tunnel between your Raspberry Pi and other devices, ensuring that all data passing through is encrypted. OpenVPN also supports different authentication methods, making it a flexible choice.
- WireGuard: Known for its simplicity and speed, WireGuard is becoming increasingly popular as a lightweight VPN solution. While still in development, WireGuard offers modern encryption algorithms and efficient performance, making it an attractive option for Raspberry Pi users.
- PiVPN: If you prefer an out-of-the-box solution, PiVPN is a fantastic choice. Built specifically for the Raspberry Pi, PiVPN simplifies the process of setting up a VPN server. It uses OpenVPN under the hood, providing a user-friendly interface and a simple installation process.

Encryption of external memory. Storage devices, such as USB drives or external hard drives, encrypting them is essential to protecting your data in the event of loss or theft. The Raspberry Pi can be used to encrypt these devices, ensuring that your files remain secure even if they fall into the wrong hands. Here are two available options:

- LUKS: Linux Unified Key Setup (LUKS) is a disk encryption specification widely used in Linux systems. The Raspberry Pi, being a Linux-based computer, can easily encrypt external storage devices using LUKS. Once encrypted, devices can only be accessed with the correct password or key, providing an extra layer of security.
- VeraCrypt: As mentioned earlier, VeraCrypt is a versatile encryption software that supports external storage devices. By creating an encrypted vault on your USB drive or external hard drive, you can ensure that your data remains protected even if the device is lost or stolen.

Encrypting data using the Raspberry Pi is an effective way to improve security and privacy. From using encryption software to setting up a VPN or encrypting external storage, there are different options available to suit different needs. While the best option may vary depending on individual requirements, VeraCrypt stands out as a versatile choice for encrypting files and containers. When it comes to VPN solutions, OpenVPN and PiVPN offer strong security features and ease of use. Finally, for external storage encryption, LUKS and VeraCrypt provide robust encryption capabilities. Using these techniques, you can protect your files and communications with the Raspberry Pi, ensuring that your data remains secure.

7 Adding an Additional Layer of Security

Implementing two-factor authentication (2FA) on the Raspberry Pi means adding an extra layer of security of robust security measures. By adding an extra layer of security, 2FA ensures that only authorized people can access your device and its resources. In this section, we will explore the benefits of implementing 2FA on the Raspberry Pi, discuss different methods for enabling it, and compare the options to determine the best solution. Namely, the advantages of implementing 2FA on Raspberry Pi are the following:

- Enhanced security: Two-factor authentication adds an extra layer of security beyond a password. It requires another form of authentication, such as a physical token or one-time password (OTP), making it difficult for unauthorized persons to gain access.
- Protection against password-based attacks: Traditional password-based authentication methods are vulnerable to brute-force attacks and password guessing. By implementing 2FA, even if an attacker manages to obtain the password, they will still need another factor to gain access.
- Mitigating phishing attacks: Phishing attacks, where attackers trick users into revealing their credentials, are prevalent. With 2FA, even if the user falls victim to a phishing attack, the attacker would still not be able to access the Raspberry Pi without another authentication factor.

On the other side, the methods that enable 2FA on Raspberry Pi are the following:

- Time-Based One-Time Password (TOTP): TOTP is a widely used method for implementing 2FA. It generates a unique one-time password that changes every few seconds, which the user needs to enter along with their password. This can be achieved by using mobile apps like google Authenticator or Authy, which generate an OTP based on a shared secret key.
- Hardware tokens: Hardware tokens, such as YubiKeys, provide an additional level of security by generating OTPs. These physical devices connect to your Raspberry Pi via USB and require the user to tap a button or touch the device to generate an OTP. Hardware tokens are very secure and convenient, as they eliminate the need for a mobile app.
- Biometric authentication: The Raspberry Pi can also use biometric authentication methods, such as fingerprint scanners or facial recognition, to implement 2FA. Biometric data is unique to each individual, making it a highly secure form of authentication. However, additional hardware and software tuning may be required.

The TOTP is a popular and practical method, as it only requires a mobile application and does not rely on additional hardware. It provides a good balance between security and usability. Hardware tokens offer the highest level of security, as they are immune to online attacks and identity theft attempts. However, they may require an upfront investment in purchasing hardware. Biometric authentication provides a seamless user experience and excellent security. However, this may require additional hardware and software tuning, while there are potential privacy issues associated with storing biometric data. Considering the options, TOTP using mobile apps like Google Authenticator or Authy is often the best choice for implementing 2FA on the Raspberry Pi. It strikes a balance between security, convenience, and cost-effectiveness. However, if you prioritize maximum security and are willing to invest in hardware tokens, they offer a great alternative. Biometric authentication can be explored if you have the necessary hardware and are comfortable with the associated privacy considerations.

8 An Example of Using Raspberry Pi for Monitoring Sensitive Cargo at Sea

We decided to conduct our own experiment because there was a dearth of information regarding the tracking of containers containing sensitive and dangerous materials (particularly nuclear material) in maritime traffic. Therefore, we proposes a practical Do It Yourself (DIY) approach for building a real-time tracking device for hazardous cargo [6–10]. This strategy aims to address the ongoing search for different solutions to problems of this nature, which can be costly and potentially hinder safe transportation.

The proposed device is made up of a Raspberry Pi 4 model (6 GB RAM, 64 GB storage capacity) and a lithium-ion battery power source that can provide a UPS for the Raspberry Pi. The Raspberry Pi functions as a central computer with connections to a number of sensors, including the crucial GSM and GPS modems as well as temperature, humidity, tilt, and light sensors (Fig. 1).

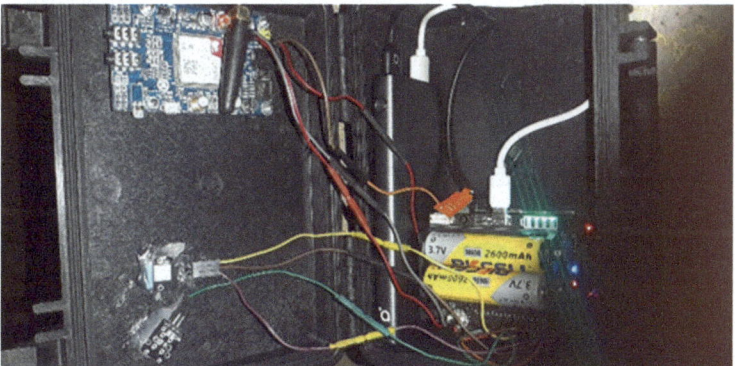

Fig. 1 Within the real-time tracking apparatus (own research)

The server side is where data gathering, processing, alarm setup, and triggering are organized. The following sensors are connected to the main computer, i.e., a Raspberry Pi for scanning temperature, humidity, lightness, tilt or motion, and time. The parts that are needed are affordable and widely accessible. These components function as an effective connection in both virtual and real environments because of the applied physical computing concept.

The sensor limits were set for the experiment's purposes. Namely, if the container deviates from its perpendicular position by more than 15°, the tilt sensor is preconfigured to sound the alarm. The humidity and temperature sensors are programmed to continuously monitor the humidity and temperature inside the container. The sensors transmit data in degrees Celsius and percentages to the server. The system sounds a light alarm if the container is opened while being transported. Lux is used as a unit for light sensors. It is considered that the container is closed and has a dark interior while it is being transported. The alarm is set off by an unintentional opening of the container door while traveling. Generally speaking, the alarms alert the dispatcher to unwanted changes in the sensor's parameters. The instrument measures 16 cm in length, 12 cm in width, and 6 cm in height. It is installed inside the door of the container to minimize interference and prevent damage during the journey (Fig. 2).

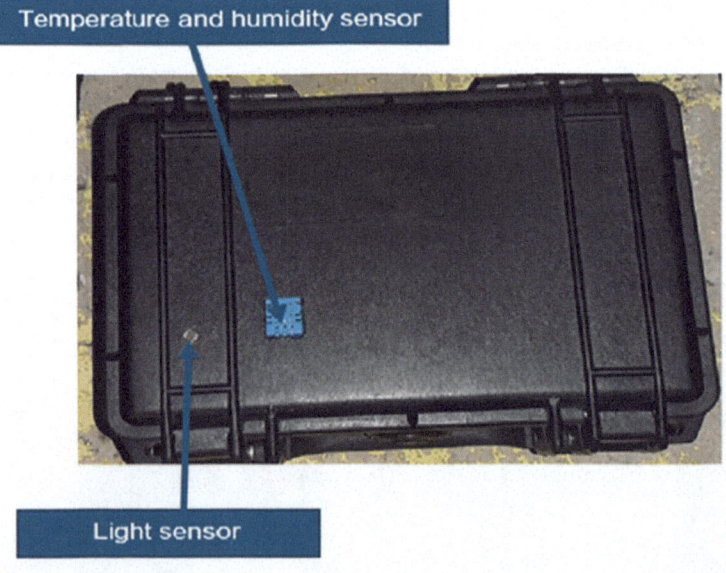

Fig. 2 Gadget from the exterior (own research)

9 The DIY Experimental Results

The digital stamp, geolocation, temperature, humidity, lightness, inclination or tilt, and time data are all gathered during the experiment. The instrument worked well. It was tested by opening and closing the container the gadget was placed in. We intentionally set off alarms in order to test the device in real time and record the opening and closing times. When loading and unloading, the tilt sensor responded. It transmitted an output of "zero" if the tilt limit was exceeded. If not, an output of "one" was sent. Every piece of recorded data has a time stamp.

As previously stated, the container was moved from the Port of Bar in the Republic of Montenegro to the Italian border via fishing boat. The container was situated on the deck. We were able to shed light on the signal's quality as well as the precision of the container's location and its load's status.

Figures 3, 4, 5, and 6 show the data that the sensors along the route collected.

The device has a background information communication system and a protected graphical interface. Figure 7 shows the path of the fishing boat carrying a container from the Port of Bar (Montenegro) to the middle of the Adriatic Sea, where the fishing boat stopped and returned to the port of origin after fishing. The application for integrating the software data and displaying it on the map is made in Paython. The device was previously tested in road traffic on the Subotica—Belgrade (Serbia) and Belgrade—Bar (Serbia—Montenegro) routes. The results were also good. This device has recently been patented in Montenegro (Patent No: P-2023-94; G-06Q10/083).

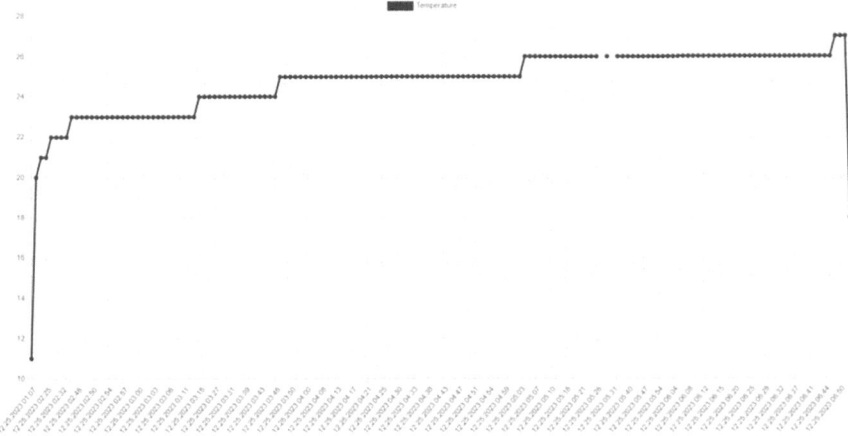

Fig. 3 A sample of data from a temperature sensor (own research)

Fig. 4 A sample of data from a humidity sensor (own research)

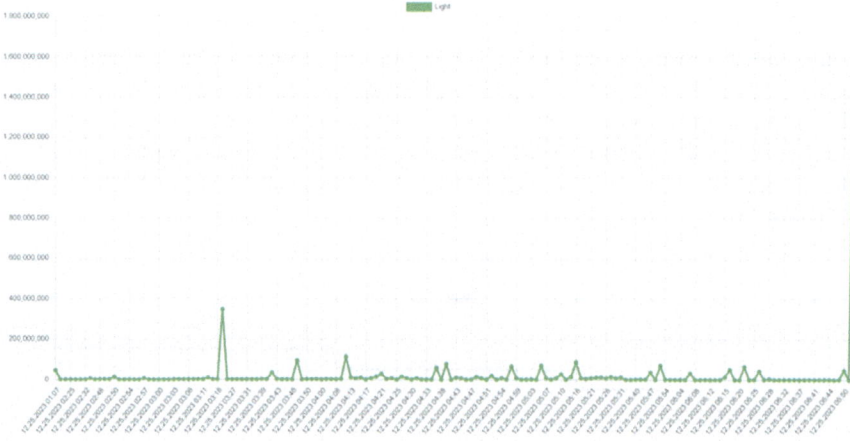

Fig. 5 A sample of data from a light sensor (own research)

10 Raspberry Pi as a Security System

Let us assume we would like to protect data on tracking and tracing cargo container with sensitive cargo. Then, when it comes to improving security and privacy, the Raspberry Pi has proven to be an incredibly versatile tool. One of its most popular applications is as a security system, capable of monitoring and detecting intrusions. In this section, we will cover the different aspects of using the Raspberry Pi as a security system, exploring its capabilities, options, and best practices for setting up an effective monitoring and intrusion detection system. The first step in setting up a Raspberry Pi security system is choosing the right model. Raspberry Pi offers a range of models, each with different specifications and capabilities.

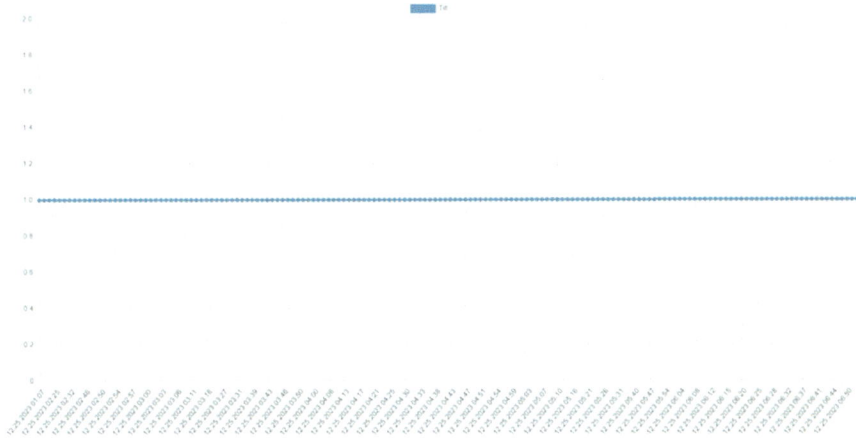

Fig. 6 A sample of data from a tilt sensor (own research)

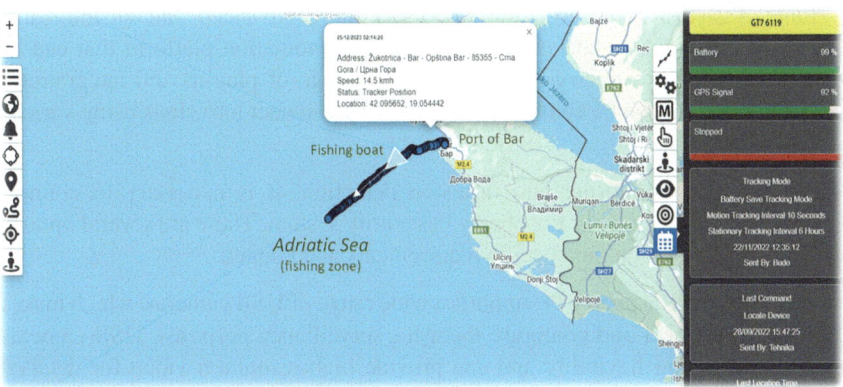

Fig. 7 Web application for tracking the container at sea: A dispatcher's view (own research)

When it comes to security systems, it is crucial to choose a model that can handle the necessary processing power and storage requirements. We can distinguish the following models:

- Raspberry Pi 4: The latest model in the Raspberry Pi line, the Pi 4 offers significant improvements in performance and memory. With its quad-core processor and up to 8 GB of RAM, it is well suited for running resource-intensive security applications.
- Raspberry Pi 3: Although not as powerful as the Pi 4, the Pi 3 is still a capable option for a security system. It offers a quad-core processor and up to 1 GB of RAM, making it suitable for most surveillance and intrusion detection tasks.

- Raspberry Pi Zero W: If you have limited space or budget constraints, the Raspberry Pi Zero W can be a viable option. Despite its compact size, it still packs a lot of punch with a single-core processor and wireless connectivity.

Once we have selected the appropriate Raspberry Pi model, the next step is to install the operating system and security software. There are several options available, each with its own advantages and characteristics:

- Raspbian: The official operating system for the Raspberry Pi, Raspbian is a Debian-based distribution that offers a user-friendly interface and comprehensive software support. It is an excellent choice for beginners and provides a stable foundation for building a security system.
- MotionEyeOS: If you're looking for a lightweight and dedicated option, MotionEyeOS is worth considering. It turns your Raspberry Pi into a network video recorder, capable of recording and analyzing video streams from connected cameras. With its intuitive web interface, MotionEyeOS simplifies the setup and management of the surveillance system.
- Home Assistant: For those looking for a more advanced and customizable solution, Home Assistant is a powerful home automation platform that can be integrated with Raspberry Pi. With its wide range of plugins and integrations, Home Assistant can be configured to monitor and detect intrusions using various sensors and devices.

For effective monitoring and intrusion detection, it is necessary to connect cameras and sensors to your Raspberry Pi security system. There are several options available, depending on our specific requirements and budget:

- USB cameras: Raspberry Pi supports a wide range of USB cameras, which makes it easy to connect and configure them for surveillance purposes. USB cameras offer positioning flexibility and can provide high-resolution video for detailed monitoring.
- Raspberry Pi camera module: Designed specifically for the Raspberry Pi, the official camera module offers excellent image quality and seamless integration. It connects directly to the camera port of the Raspberry Pi, making it a compact and convenient option for surveillance applications.
- PIR motion sensors: Pairing cameras with passive infrared (PIR) motion sensors can improve intrusion detection accuracy. These sensors detect changes in heat signatures, triggering cameras to capture footage when motion is detected. This combination ensures that the system records only relevant events, minimizing false alarms.

To get the most out of your Raspberry Pi security system, it is important to implement effective intrusion detection and notification algorithms like:

- OpenCV: OpenCV is a popular open-source computer vision library that can be used for real-time object detection and tracking. Using OpenCV algorithms, Raspberry Pi can analyze video feeds from connected cameras, identifying and flagging potential intrusions.

- Machine learning algorithms: These can be used for more advanced intrusion detection. By training the system on a dataset of known intrusions, the Raspberry Pi can learn to recognize patterns and anomalies, improving its detection accuracy over time.
- Push notifications: To receive instant alerts, the integration of push notification services such as Pushover or Telegram can be useful. These services can send notifications to our smartphones or other devices, ensuring that we are notified in a timely manner of any detected intrusions.

Raspberry Pi offers a flexible and cost-effective solution for monitoring and intrusion detection. By choosing the right model, installing the appropriate software, connecting cameras and sensors, while applying efficient detection algorithms, one can create a robust security system that improves privacy and protects premises and assets.

11 Conclusion

In the final part of security and privacy enhancements, we look at the Raspberry Pi's empowering capabilities when it comes to protecting our digital environment. We have explored various aspects of security and privacy, from securing our home networks to protecting our personal data. Raspberry Pi can be a great tool for setting up surveillance and monitoring systems, whether it is home security or workplace surveillance. With its compact size and low power consumption, it can easily be hidden or mounted discreetly. Additionally, several software options are available that allow us to set up motion detection, live streaming, and even facial recognition. For example, using the popular MotionEyeOS software, we can turn our Raspberry Pi into a powerful surveillance system with multiple cameras, motion alerts, and remote access capabilities. One of the most notable advantages of the Raspberry Pi is its ability to act as a firewall or network security device. Using software like pfsense or OpenWrt, one can turn the Raspberry Pi into a dedicated firewall that protects the entire network from external threats. This provides an extra layer of security by filtering incoming and outgoing traffic, blocking malicious websites, and monitoring network activity. Moreover, the Raspberry Pi's low power consumption makes it an energy-efficient option for 24/7 operation as a dedicated network security device. The versatility of the Raspberry Pi extends beyond traditional security measures. By combining with home automation platforms such as Home Assistant or OpenHAB, we can create a comprehensive security and privacy ecosystem for our smart homes. For instance, one can integrate motion sensors, door/window sensors, and security cameras with the Raspberry Pi home automation system to automate security alerts, control access to your home, and monitor activities remotely. This integration provides a holistic approach to security and privacy, making it easy to manage and control all aspects from a single platform. Another area where the Raspberry Pi shines provides secure and private data storage solutions. We can set up Raspberry Pi as a personal cloud server using software such as Nextcloud or ownCloud, which allows you to securely store and access your files from anywhere. By adding external hard drives or network attached storage, we can expand storage capacity. In addition, we can use encryption techniques to protect sensitive data stored on the Raspberry Pi, ensuring that even if the physical device falls into the wrong hands, the data remains safe. Raspberry Pi offers countless possibilities to improve security and privacy. Whether it is setting up surveillance systems, strengthening network security, or integrating home automation and privacy tools, Raspberry Pi lets users take control of their digital lives. Its affordability, flexibility, and community-driven support make it an excellent choice for those looking to strengthen their security and privacy measures.

Acknowledgments Research for this publication was funded by the EU Horizon2020 project 952360-MariCybERA.

References

1. N.X. Arreaga, G.M. Enriquez, S. Blanc, R. Estrada, Security vulnerability analysis for IoT devices Raspberry Pi using PENTEST. Proc. Comput. Sci. **224**, 223–230 (2023). https://doi.org/10.1016/j.procs.2023.09.031
2. K. Rzepka, P. Szary, K. Cabaj, W. Mazurczyk, Performance evaluation of Raspberry Pi 4 and STM32 Nucleo boards for security-related operations in IoT environments. Comput. Netw. **242**, 110252 (2024). https://doi.org/10.1016/j.comnet.2024.110252
3. R.A. Nadafa, S.M. Hatturea, V.M. Bonala, S.P. Naikb, Home security against human intrusion using Raspberry Pi. Proc. Comput. Sci. **167**, 1811–1820 (2020). https://doi.org/10.1016/j.procs.2020.03.200
4. S.E. Mathe, H.K. Kondaveeti, S. Vappangi, S.D. Vanambathina, N.K. Kumaravelu, A comprehensive review on applications of Raspberry Pi. Comput Sci Rev **52**, 100636 (2024). https://doi.org/10.1016/j.cosrev.2024.100636
5. M. Tanque, P. Bradford, Chapter six—Virtual Raspberry Pi-s with blockchain and cybersecurity applications, in *Advances in Computers*, ed. by A.R. Hurson, vol. 131, (Elsevier, 2023), pp. 201–232. https://doi.org/10.1016/bs.adcom.2023.04.005
6. S. Bauk, R. Dzankic, Smart cargo container tracking and IT security management: experimental results. J. Marit. Res. **21**(2), 128–134 (2024) https://www.jmr.unican.es/index.php/jmr/article/view/851/855
7. S. Bauk, Modelling radioactive materials tracking in sea transportation by RFID technology. TransNav **14**(4), 1009–1014 (2020)
8. S. Bauk, R. Dzankic, Model of tracking radioactive cargo in sea transport, in *The Proceedings of the 10th Mediterranean Conference on Embedded Computing (MECO'2021), Budva, Montenegro, 7–10 June 2021*
9. S. Bauk, A. Radulovic, R. Dzankic, Physical computing in a freight container tracking: an experiment, in *The Proceedings of the 12th Mediterranean Conference on Embedded Computing (MECO'2023), Budva, Montenegro, 6–10 June 2023*
10. R. Džankić, S. Bauk, Tracking hazardous cargo at sea: an experiment, in *Proc. of the 16th ITeO 2024*, (APEIRON University, Banja Luka, Republic of Srpska, 2024), pp. 27–35, https://iteo.rs.ba/sites/default/files/ITeO_Zbornik%20radova%202024%20%281%29.pdf

Open Access This chapter is licensed under the terms of the Creative Commons Attribution-NonCommercial-NoDerivatives 4.0 International License (http://creativecommons.org/licenses/by-nc-nd/4.0/), which permits any noncommercial use, sharing, distribution and reproduction in any medium or format, as long as you give appropriate credit to the original author(s) and the source, provide a link to the Creative Commons license and indicate if you modified the licensed material. You do not have permission under this license to share adapted material derived from this chapter or parts of it.

The images or other third party material in this chapter are included in the chapter's Creative Commons license, unless indicated otherwise in a credit line to the material. If material is not included in the chapter's Creative Commons license and your intended use is not permitted by statutory regulation or exceeds the permitted use, you will need to obtain permission directly from the copyright holder.

Cybersecurity and Commercial Shipping: Is There a Need for Unification?

Ann Fenech and Sebastien Lootgieter

1 Introduction

Maritime lawyers normally are neither IT lawyers nor are they AI lawyers or experts on digitalization and information technology. We do however need to come to terms with and assist with the resolution of the diverse legal conundrums very frequently the direct fall out of cyber-attacks on the maritime space. In our work, before we find the solutions we need to understand the problem, the extent of it, and how it manifests itself. There are fairly mind-blowing statistics available which really underline the not too rosy scenario related to cybersecurity attacks on shipping which at the very least get us to address the issue and try to find a way out of what can turn into a cyber maze.

According to the British Ports Association, cyber-attacks in the maritime transport sector have increased fourfold since February 2020 and there is very little doubt that cyber-attacks aimed at vessels, port infrastructures and port administrations, international shipping organizations, flag administrations, and classifications societies are not only a direct attack on the safety of navigation and thus a direct threat

---

This chapter was originally presented as a paper by Ann Fenech at the Maritime Traffic and Cyber Security Series Digital Mare Nostrum on the 2 June 2024.

A. Fenech (✉)
President, Comite Maritime International and Partner at Fenech & Fenech Advocates Malta, Belt Valletta, Malta
e-mail: ann.fenech@fenechlaw.com

S. Lootgieter
CMI International Working Group on Cyber Security and Avocat at SCP Villeneau Rohart Simon and Associes, Paris, France

© The Author(s) 2025
S. Bauk (ed.), *Maritime Cybersecurity*, Signals and Communication Technology, https://doi.org/10.1007/978-3-031-87290-7_13

to the lives of the seafarers on board but a direct threat to international trade, the environment [1] and thus world order.

A report in 2019 from a Singaporean cyber risk management company suggested that a ransom wear attack on Asian ports could cost the global economy as much as USD 110 billion.

The mind boggles.

2 The Comite Maritime International

How does the Comite Maritime International feature in all of this? The CMI is a nongovernmental organization established in 1897.

"Unification of International Maritime law," was, is, and will remain, at the very heart of and is the very raison d'etre of the Comite.

Albert Lilar President of the Comite from 1947 to 1976[1] has stated:

> The history of maritime law bears the stamp of a constant search for stability and security in the relations between the men who commit themselves and their belongings to the capricious and in dominatable sea. Since time immemorial the postulate which has inspired all the approaches to the problem has been the establishment of a uniform law.

Albert Lilar and Carlo Van Den Bosch in their history of the CMI[2] in turn quote Louis Frank one of the co-founders of the Comite in 1897 as having stated:

> It is with the object of overcoming multiple opposition, national particularism, of resolving difficulties not by means of abstract and theoretical solutions but from the needs of practice, of obtaining the ear of the Parliaments, that we had the idea of appealing, not only to the jurists who are interested in maritime law, but to the very people who, in their interests, in their problems of every day, have to submit to the consequence of good and bad laws. We have taken into consideration that the shipowner, the merchant, the underwriter, the average adjuster, the banker, the person who is directly interested, all take a preponderant part in our work; that the task of the jurist is to discern that which, among these divergent interests, is common to all; to discern what, among the diverse solutions, is the best to contribute one's learning and one's experience; but that in the final analysis, the jurist must hold the pen and that it is the man with the experience who must dictate the solution.

This quote holds good today over 130 years later. The modus operandi of the Comite Maritime International in identifying problems and challenges, in forming international working groups made up of maritime lawyers from various jurisdictions and persons directly from the industry ensures that the solutions it provides and suggests are solutions which have been designed ultimately by as Mr. Franck put it so many years ago "the man with the experience who must dictate the solution."

---

[1] Albert Lilar and Carlo Van Den Bosch "Le comite maritime international 1897–1972" p 2.
[2] Ibid p 12.

This has been a very successful formula leading the Comite Maritime International to be responsible for the initial drafting of the vast majority of international maritime Conventions. The success behind the drafting initiative stems from the modus operandi of the CMI. When the CMI is persuaded that it may be useful to research into the need or otherwise of having a convention on any particular subject, a working group is formed to consider the matter and very frequently the working group will put together a questionnaire which is then sent to all the national maritime law associations with a view to seeing and understanding if and how the domestic law of those countries deal with the subject. Thus, at the very start of the project, the CMI is usually in possession of very useful information on how or whether the subject matter is dealt within the individual and diverse legal systems. Today we have 57 national maritime law associations, members of the CMI—a far cry from the original 8 founding members of the CMI in 1897 which were Belgium, Denmark, France, Germany, Italy, The Netherlands, Norway, and the United Kingdom.[3]

Why should an organization spend over 125 years attempting to unify the maritime laws of different nations. Well, shipping by its very nature is one of the most international of areas of activity, responsible for the carriage of 90% of world trade. Put very simply we would not have international trade, the food on our table or the clothes we wear or the energy for our industries and homes, if it were not up to these magnificent objects which carry the goods from one country to another. Our very existence depends on them.

Precisely because of the internationality of the activity, a typical maritime scenario could include the application of literally dozens of competing jurisdictions and laws. The law of the owner of the vessel could be different to the law of the charterer, to the laws of the thousands of owners of the containers on board, to the law of the place where the incident occurs, to the law of the flag of the vessel, to the law of the flag of any other vessel involved in a casualty, and the list goes on.

Furthermore, each stakeholder could have their respective contractual arrangements.

And that is precisely the role of the CMI which is to draft international legislation seeking to unify international maritime law so that as far as possible the dynamics related to the maritime space are regulated by the same law bringing about uniformity and bringing about stability. Since 1897, the CMI has been responsible for drafting the vast majority of international maritime conventions. These include conventions on carriage of goods, collisions, salvage, limitation of liability, arrest of ships, responsibility for pollution, etc. and are constantly working on new projects. The burning question is—is there room for an international convention to ensure cybersecurity?

---

[3] Frank Wiswal, A brief history CMI. Website a-brief-history-wiswall.pdf (comitemaritime.org).

3 Cybersecurity: A Major Risk

Maritime traffic is extensively dependent on information systems and computer networks which are an intrinsic part of navigation, cargo loading management systems, general vessel management systems, port managements systems, flag administrations, and classifications societies, leading one to the conclusion that it is impossible for ships to operate without reliable information technologies through extensive computer, IT and AI networks.

Just considering this very simple statement leaves the hairs of your body stand on edge when one considers the significant security threat to the entire shipping world and international trade in the event of a cyber-attack on a vessel's systems.

The growing threats to the safety and security of a vessel, of a voyage, and of shipping in general are significant. Different events have varying effects on the frequency and severity of cyber-attacks—geopolitics, regional wars, international terrorism, sanctions busting, pandemics, and digitalization. According to a cybersecurity consulting firm Naval Dome "attempts of maritime cyberattacks have increased by 400% during the COVID-19 crisis."

As stated, cyber-attacks are not limited to attacks on a single vessel or ship owner or ship manager but we have seen attacks say on the entire information platforms of port authorities or of international organizations which handle very important data on ships such as, for instance, the attacks on the Port of Antwerp, on the Port of Durban, entire flag administrations, a major classification society, and even the IMO.[4]

Cyber-attacks come in all forms from the most basic to the most sophisticated including:

- The theft of sensitive information.
- Intrusion into the control system of a port.
- Phishing, with which hackers extract personal information from their victims by pretending to be a bank or an administration.
- The spread of false information about a company.
- Embezzlement.
- Ransom demands after a hacker blocks a company's computer system.
- The intrusion into the navigation system with a view to causing damage.
- The intrusion into the loading system of a vessel.

The effects of these attacks are often very significant. They may have an effect not only on the vessel itself, directly impacting the safety and security of the vessel and of all her crew members, or on the company owning the vessel but may have serious effects on world trade generally.

Imagine a cyber-attack on a vessel passing through the Suez Canal and into the Mediterranean or while in the straits of Gibraltar which would paralyze her

---

[4] IMO security breached by 'sophisticated' cyber attack: Lloyd's List. Available at: https://lloydslist.com/LL1134099/IMO-security-breached-by-sophisticated-cyber-attack

navigation system while in the straits or the canal effectively blocking and closing down these vitally important routes to international trade.

While this attack has not yet occurred, we have already seen what the effect of an immobilized vessel in the canal actually means. On 23 March 2023, the Ever Given which was on her way to Rotterdam, grounded and blocked the Suez Canal for almost a week essentially causing the delay of 422 ships, carrying 2 million tons of goods immobilizing 10% of global trade with a huge effect on the transport of petroleum causing oil prices to rise.

For nearly 9 months now, the Houtis have randomly been attacking vessels. The events surrounding the Greek flagged tanker the MV Sounion carrying 150,000 tons of crude oil are well documented.[5] We now know the results with numerous shipping lines having taken the decision of avoiding the Red Sea area completely, thus giving up on a Suez transit and are opting for going around the Cape instead of passing through the Mediterranean. This is adding an extra 3 weeks on to the voyage from the Far East to Europe and increasing the costs of such a voyage significantly in hire per day and in additional bunkers that need to be consumed.

We also saw very recently what happens when a vessel appears to be proceeding perfectly normally through a busy waterway suddenly clearly loses control over her navigation systems and collides straight into a bridge such as what occurred in relation to the vessel Dali and the Baltimore Bridge on 26 March 2024. This led to loss of life, very serious damages to the bridge, closure of main highways, blockage of the entire port area, and delays in shipping containers. The list is endless. Numerous claims have followed including a claim from the US government seeking over USD 100 million in costs and damages filed on 18 September 2024 which consider that the Dali was "abjectly unseaworthy." In the initial stages, the press did identify a cyber-attack on the vessels navigation system as possibly responsible however this has as yet not been established.

If one were to dig just a little bit deeper, in the event that a cyber-attack on a vessel's navigation system leads to a collision or a grounding, that would almost certainly lead to damages caused not only to the vessel itself but to third parties. In such a case who would be liable to the third parties? Is it the ship owner, the classification society, the software designer, the perpetrator? Is there any particular legal regime to protect an innocent third party from cybersecurity incidents.

If a vessel's cargo loading operation system is hacked leading to damages suffered to the cargo leading in turn to delays in the voyage, are the owners liable for such damage?

A question that will arise and is arising now is how far do we take the obligation on the ship owner to provide a seaworthy ship in the context of the due diligence exercises which need to be carried out on a vessel's preparedness for cyber-attacks? Does the ship owner have an obligation to ensure that his ship has the best form of cyber protection available or does it extend to ensuring that the most thorough and rigorous assessment of every single IT system on board is carried out? What is the

---

[5] Web sites: www.cnn.com; www.bbc.com; www.theguardian.com

standard set failing which will be tantamount to fault or negligence on the part of the shipowner? Should the level of preparedness of the crew and the training of the crew in dealing with cyber-attacks be taken as being part of the ship owners obligation to provide a seaworthy vessel?

3.1 The Hacking of the Port of Antwerp in 2011

The hacking of the Port of Antwerp in 2011, as reported by Stormshield, is described as follows:[6]

"In 2013, the port of Antwerp discovered that a drug cartel had hijacked its container management system. In fact, the port's computer network had been spied on since June 2011, when the network was reportedly infiltrated by malware, specifically a keylogger (which allowed the hackers to record the keystrokes used by the loading/unloading operators, and thereby obtain usernames and passwords).

The port of Antwerp eventually re-secured its system by investing nearly €200,000 to set up countermeasures, including a new password management system (to provide access to containers) and new communication channels between port operators and customer services."

In the end, the drug cartel was able to know the location where the containers were stored, their identification numbers, and the passwords to take possession of them."

3.2 Impact of the 2017 NotPetya Cyber-Attack on Maersk

In 2017, shipping giant Maersk, responsible for transporting over 12 million containers worldwide, suffered a significant cyber-attack. Wired provides a detailed account of the events as they unfolded.

> All across Maersk headquarters, the full scale of the crisis was starting to become clear. Within half an hour, Maersk employees were running down hallways, yelling to their colleagues to turn off computers or disconnect them from Maersk's network before the malicious software could infect them...Disconnecting Maersk's entire global network took the company's IT staff more than two panicky hours...The malware's goal was purely destructive. It irreversibly encrypted computers' master boot records, the deep-seated part of a machine that tells it where to find its own operating system. Any ransom payment that victims tried to make was futile. No key even existed to reorder the scrambled noise of their computer's contents. The release of NotPetya was an act of cyberwar by almost any definition—one that was likely more explosive than even its creators intended. Within hours of its first appearance, the worm raced beyond Ukraine and out to countless machines

---

[6] Nicaise, V. (2022) Port Cyberattack: Hackers & Maritime Cybersecurity, Stormshield. Available at: https://www.stormshield.com/news/cybermaretique-a-short-history-of-cyberattacks-against-ports/

around the world, from hospitals in Pennsylvania to a chocolate factory in Tasmania. It crippled multinational companies including Maersk.[7]

The financial impact of NotPetya for Maesrk reportedly ran into the USD 300,000,000.

According to Trade winds: "Maritime companies have admitted paying ransomware demands in increasing numbers, with the average cost of unlocking computer systems reaching $3.2m in 2023, according to a new study. A survey of more than 150 industry professionals in the maritime sector found that 14% admitted to paying a ransom following a cyber-attack in 2023 compared with 3% the previous year, according to the study by law firm HFW and maritime cyber security company CyberOwl."[8]

A report in 2019 from a Singaporean cyber risk management company suggested that a ransomwear attack on Asian ports could cost the global economy as much as USD 110 billion.

3.3 AIS Spoofing

In 2023, an increase in the imposition of sanctions has seen criminal ship owners who are part of what is termed as the "dark fleet" conduct AIS spoofing.[9] Today, every commercial vessel is obliged to carry an AIS—automatic identification system so that its location can be tracked on a 24/7 basis. From our smart phones and for those who have the AIS app, we can input the name of a vessel and find out exactly where she is. This is of course bad news for these dark fleet operators who wish to breach sanctions by carrying illegal cargoes and trades. So what do they do?

They manipulate the information systems by succeeding in changing their location on AIS. There are dozens of instances where AIS would show that vessels are off West Africa when in fact, they would be sanction busting off the coast of Venezuela. Or they would be showing as being off the coast of India when they are really loading sanctioned oil in Iran. They would have the appearance of being at anchor, indicating the ship may be used for offshore storage. However, a thorough analysis of the vessel's signal activity, along with human or imagery sources would confirm the ship's "transmitted" location to be inaccurate.

---

[7] Excerpt, A.G. (2018) The untold story of notpetya, the most devastating cyberattack in history, Wired. Available at: https://www.wired.com/story/notpetya-cyberattack-ukraine-russia-code-crashed-the-world/

[8] Shipping names pay multimillion-dollar ransoms after cyber-attacks, TradeWinds. Available at: https://www.tradewindsnews.com/technology/shipping-names-pay-multimillion-dollar-ransoms-after-cyber-attacks/2-1-1536556

[9] M. Wiese Bockmann, An accident waiting to happen. How the 'dark' fleet of tankers shipping sanctioned oil uses deceptive shipping practices and regulatory arbitrage to avoid scrutiny, Presentation made at CMI Colloquium 2024 in Gothenburg, Sweden, https://comitemaritime.org/presentations

This can be established through specialized governmental naval assets and through the engaging of professional organizations who specialize in this activity and who have the tools to actually find out when there are suspicions, that a particular vessel is engaged in AIS spoofing. Much of the investigation however remains dependent on visual sightings from human sources and imagery which would indicate that this data is inaccurate, and the vessel is, in fact, operating out of sanctioned ports or lifting sanctioned cargoes.

3.4 Automation

No discussion on the subject would be complete without touching upon the development of unmanned ships and the increased risk of cyber-attacks.

I cannot but agree with the view that the more we advance in the use of artificial intelligence for the purposes of vessel management the more alert we have to be to the risk in the rise of cyber-attacks and the effect those attacks would have. I speak here principally of MASS, Maritime Autonomous Surface Ships, often referred to as Unmanned ships.

We have currently several categories and automation levels of MASS ranging from AL0—Manual Control; to AL1—On board decision support; AL2—on board and off shore decision support; AL3—Active human in the loop who monitors and authorizes; AL4—Human in the loop where decisions and actions are performed autonomously with human supervision and where high impact decisions give human operators the opportunity and override; AL5 is monitored fully autonomous meaning rarely supervised operations where decisions are entirely made and actioned by the system; AL6—Fully autonomous, unsupervised operation where decisions are entirely made and actioned by the system.

So, we see here a huge range from total manual human control to the complete absence of all human control.

One doesn't have to be a rocket scientist to figure out that the higher the level of autonomy and dependency on artificial intelligence, and the more reduced presence of the human being in the context of maritime operations, the more room and scope there is going to be for cyber-attacks on the vessel's systems and the systems of the computer companies which are going to be responsible for the systems on board and on shore.

As referred to earlier members of the maritime industry work under specific contractual arrangements. They have their general conditions, their charter-parties, and their insurance policies. Some of these forms address cyber risk, other remain silent. Take the example of insurance policies, Lloyd's of London have excluded cyber risks from insurance cover unless there is a specific agreement between the parties.

4 Should Cybersecurity Be Regulated?

4.1 The Current Norms

In the context of this discussion on maritime traffic and cybersecurity, is there an international regulatory legal order which attempts to regulate this space?

- ***IACS**—the International Association of Classification Societies* published the Unified Requirement (UR) E26 in April 2022, which brings together a number of requirements. This started to apply to all ships with construction contracts signed from January 1, 2024.
- ***BIMCO, CLIA, ICS, INTERCARGO, INTERTANKO*** have all issued guidelines on cybersecurity and cyber safety on board ships [2].

  The International Convention on Cybercrime adopted in Budapest on November 23, 2001, enumerates a number of offenses related to cybercrime (for example, illegal access, damage to system and data integrity, forgery, etc.) and requires each contracting state to adopt the necessary laws to make these acts criminal offenses in their legal system.
- ***The EU issued a Directive No. 2016/1148 of July 6, 2016***, concerning measures to ensure a high common level of security of network and information systems across the European Union (known as the NIS Directive, Network and Information System Security).
- ***The IMO joint Maritime Safety Committee and Facilitation Committee*** have provided important Guidelines on Maritime Cyber Risk Management that can be incorporated into existing risk management processes. These guidelines recommend all entities acting in the maritime transport sector to adopt risk management systems consistent with the current safety and security management shipping practices. They recommend that the cyber risk management should consider several types of control, management, operational, procedural, and technical processes. In 2017, the same MSC adopted resolution MSC.428(98) on Maritime Cyber Risk Management in Safety Management Systems (SMS) which encourages flag state administrations to review the maritime cybersecurity in accordance with the objectives and requirements of the ISM code.

While all of this is very good indeed is it enough? Clearly not and is there anything else we can do to better the situation and address these serious difficulties? Will these difficulties be resolved by the creation of a new international Convention on cybersecurity?

4.2 The Role of the CMI

In view of all of this, the CMI decided to do what it usually does when it sees that a maritime situation may benefit from uniform rules. It set up an International

Working Group made up of practicing lawyers from various parts of the world with the brief of studying the situation and making recommendations. This group was set up and first started its debates at the CMI conference in New York in 2016. At the time, it was difficult to precisely assess such risk and what the CMI could do. Members of the IWG mostly had in mind insurance issues. However, it is not the CMI's job to draft insurance policies and clauses. With the development of digitalisation and further threats to world order, we all started to understand better the magnitude of the problem and decided that perhaps a starting point would be:

1. To raise a type of "cyber awareness" among the stakeholders via the national maritime law associations about cybersecurity in particular in jurisdictions where it might be overlooked by local regulatory powers or in jurisdictions which are not yet prepared for dealing with these eventualities.
2. To check whether the risk of cybersecurity would lead to the amendments of existing international conventions, in particular SOLAS and ISPS.

In order to pursue these goals, the IWG has now drafted an extensive questionnaire which once approved will be sent to each and every national maritime law association member of the CMI.

The draft questionnaire as it currently stands includes questions from the most basic to the most complex and includes questions such as:

To what extent do you believe that existing international conventions do not adequately cover cyber risk, do you believe this would best be rectified by:

1. A new bespoke cyber risk convention.
2. Amendment to various existing international conventions.
3. An addition to the Safety of Life at Sea Convention covering cyber risk.

And questions geared to gathering statistics on the amount and type of cyber-attacks in the respective jurisdictions on ship owners, ship managers, vessels, port authorities, and organizations in the logistics supply chain.

The plan is to finalize this questionnaire soon and get it sent out by the third quarter of the year.

Once all the replies are in, the CMI will be in a better position to assess whether an international convention on the matter would be useful and if the results indicate that it will be useful, then the CMI will put its thinking cap on and start drafting.

5 Conclusions

The entire object of this paper was to provide some food for thought. Maritime cybersecurity has certainly improved, but it is still considered that the maritime industry remains an easy target. Implementing cybersecurity measures for an entire fleet is not a simple process and due consideration must be given to the peculiarities of each and every ship.

The effects of cyber-attacks on shipping go to the very heart of what we must all strive to achieve—safer shipping for crews on board, cleaner seas and oceans, the smooth operation of international trade, and generally a safer environment.

Cyber-attacks on shipping is an extremely dangerous space which, with the on boarding of artificial intelligence for the purposes of operating ships generally, is becoming even more challenging. For this reason alone, the laws of different states must sing from the same hymn sheet, must have a similar regulatory framework, and must provide systems where the perpetrators of these attacks know that when they are caught there is absolutely zero tolerance to what ultimately is pretty despicable behavior which threatens the world order and the lives of people who serve as seafarers to whom we owe a great deal.

References

1. M. Wiese Bockmann, An accident waiting to happen. How the 'dark' fleet of tankers shipping sanctioned oil uses deceptive shipping practices and regulatory arbitrage to avoid scrutiny, Presentation made at CMI Colloquium 2024 in Gothenburg, Sweden, https://comitemaritime.org/presentations/
2. BIMCO, *Cyber Security Workbook for On Board Ship Use*, 6th edn. (BIMCO, 2025)

Open Access This chapter is licensed under the terms of the Creative Commons Attribution-NonCommercial-NoDerivatives 4.0 International License (http://creativecommons.org/licenses/by-nc-nd/4.0/), which permits any noncommercial use, sharing, distribution and reproduction in any medium or format, as long as you give appropriate credit to the original author(s) and the source, provide a link to the Creative Commons license and indicate if you modified the licensed material. You do not have permission under this license to share adapted material derived from this chapter or parts of it.

The images or other third party material in this chapter are included in the chapter's Creative Commons license, unless indicated otherwise in a credit line to the material. If material is not included in the chapter's Creative Commons license and your intended use is not permitted by statutory regulation or exceeds the permitted use, you will need to obtain permission directly from the copyright holder.

Index

A
Anomaly detection, 124, 127–129, 143, 148, 151, 153, 154, 198
Artificial intelligence (AI), 1–12, 15, 26, 84, 129, 235, 238, 242, 245
Artificial neural networks (ANNs), 161–177
Authentication, 108, 110–113, 115, 146, 186, 219, 222–224
Automatic identification system (AIS), 22, 85, 103–117, 123, 164, 187, 241–242

B
Behaviour change, 31, 40–42, 48

C
Commercial shipping, 235–245
Concept, 6, 10, 23, 32, 35, 44, 47, 68, 85–87, 99, 109, 110, 144, 145, 156, 180, 222, 226
Cyber, 2, 15, 30, 55, 104, 123, 144, 162, 179, 197, 235
Cyberattack, 16–21, 29–33, 35–37, 39, 41, 43, 47, 103, 104, 123, 126, 137, 144, 146, 179–181, 184, 185, 188, 191, 195–197, 199, 201, 206, 207, 209–212, 238, 241
Cyber attackers, 179, 182, 184, 187, 191
Cybersecurity, 1–12, 15–26, 29, 55–81, 103, 123, 143, 161–177, 179–191, 196, 235–245
Cybersecurity awareness, 16, 30–32, 34, 35, 37, 39, 42, 43, 46–48, 55, 64, 188
Cyber threats, 4, 15, 20–26, 29–36, 38–40, 42–46, 48, 59–61, 63–65, 74, 123, 147, 157, 161–163, 176, 179, 196, 198, 208, 220, 221

D
Dead reckoning systems, 176
Deep learning in maritime systems, 195–213
Digital age, 29–48
Digitalization, 1, 2, 16, 44, 48, 55, 56, 60, 66–70, 74, 76, 83, 84, 161, 191, 235, 238, 244
Digitalization at sea, 60

F
Fraud, 61, 62, 180

G
GPS spoofing, 36, 189, 190

I
Identity theft, 181, 190, 224
Inductive logic programming (ILP), 123–138
Integrated navigation system (INS), 86, 87, 149, 163, 197
Intrusion detection system (IDS), 80, 143–157, 162, 163, 185, 190, 228

L

Level of paranoia (LoP), 46–47
Literature review, 60, 62, 197

M

Marine radar security, 198, 207, 208, 210, 213
Maritime, 1, 15–26, 29, 55, 83, 103, 123–138, 144–145, 161–177, 179–191, 195, 225, 235
Maritime automation, 123–137
Maritime cybersecurity, 15–26, 31, 33, 35, 36, 38, 42, 44, 47–48, 64, 137, 179–191, 196, 197, 240, 243
Maritime digital transformation, 29, 44, 84
Maritime education, 23, 34, 35, 37–39, 47, 48
Maritime industry, 1, 15–26, 29–39, 41–43, 47, 48, 56, 83, 84, 88, 114, 161, 163, 191, 242
Mitigation, 91, 146
Mitigation strategies, 38, 153, 181, 190–191, 196

N

Network-based attacks, 196, 199, 213

O

1D Convolutional Neural Network (1D CNN), 196, 198, 203, 204, 206–210, 213

P

Performance, 70, 77, 84, 133, 136, 146, 153–156, 166, 168, 170, 172–174, 179, 196, 201, 206–213, 221, 223, 229

Privacy, 104, 113, 114, 145, 149, 161, 186, 217–232
Proactive measures, 155, 220, 221

R

Regulations, 9–12, 17, 21–22, 24, 25, 32, 35, 56, 64, 70, 73, 75–77, 83, 90, 104, 123, 146–147
Regulations unification, 9–12, 17, 21–22, 24, 25, 32, 35, 56, 64, 70, 73, 75–77, 83, 90, 104, 123, 146, 147
Rudder controller, 83–99

S

Seafarer education, 15, 29–48
Seaport, 55, 57, 60, 62, 84
Security, 1–12, 15, 29, 55, 104, 126, 143, 161, 179, 195, 217–232, 236
Skills development, 15–26
Social engineering, 33, 36, 44, 57, 75, 179–191
Spoofing, 36, 104, 116, 117, 123–138, 146, 176, 181, 188

U

Unmanned sea-surface vessel, 83–99

V

Vulnerabilities, 15, 17, 20, 26, 29, 30, 32–34, 36, 38, 55, 60–62, 64, 65, 104, 112, 113, 123, 126, 145, 147, 150, 155, 156, 161, 163, 176, 179, 184, 188, 190, 191, 195, 197, 218, 219, 221

The manufacturer's authorised representative in the EU is Springer Nature Customer Service Centre GmbH, Europaplatz 3, 69115 Heidelberg, Germany. If you have any concerns regarding our products, please contact ProductSafety@springernature.com

Printed and bound by CPI Group (UK) Ltd, Croydon, CR0 4YY
26/03/2026
02078979-0001